D0535840

EXPLORING REQUIREMENTS:
QUALITY BEFORE DESIGN

EXPLORING REQUIREMENTS: QUALITY BEFORE DESIGN

by
Donald C. Gause
Gerald M. Weinberg

Dorset House Publishing
353 West 12th Street
New York, NY 10014

Library of Congress Cataloging-in-Publication Data

Gause, Donald C.
    Exploring requirements : quality before design / Donald C. Gause,
Gerald M. Weinberg.
      p.   cm.
    Includes bibliographical references.
    ISBN 0-932633-13-7 :
    1. Design, Industrial. 2. New products.   I. Weinberg, Gerald M.
II. Title.
    TS171.G38   1989
    745.2—dc20   89-40440
                                                          CIP

Cover Design: Marc Rubin, Rubin Design
Illustrations in Figs. 2-1, 2-2, 2-3, 7-1, 11-1, 11-4, 14-1, 14-2, 14-3, 14-4, 23-1 and 25-3: Marc Rubin, Rubin Design

Printed in the United States of America

Library of Congress Catalog Card Number 89-40440
ISBN: 0-932633-13-7   24 23 22 21 20 19 18 17 16 15 14

*To the thousands of product and systems designers worldwide
whose ingeniously clever, elegantly conceived solutions
so often amaze us by solving the right problems,
we dedicate this book.
We hope it meets your requirements.*

Acknowledgments

This work represents a collection of ideas that have been developed, refined, and tested through more than sixty total years of consultations and workshops with such organizations as IBM, Pacific Telesis, McDonnell Douglas, Philips, and Ericsson, as well as many smaller companies, less well-known but also vigorous. We express our appreciation to these organizations for encouraging our work.

We also appreciate the special efforts of Patty Terrien for following the first rough notes, asking critical questions, and providing background material; Ken de Lavigne and Eric Minch for disambiguating some of our greatest ambiguities; Stiles Roberts, Jim Wessel, and Janice Wormington for their diligence and suggestions on our first drafts; and Marty Fisher and Howie Roth for providing a steady stream of *real* designers with *real* problems. Very special thanks go to Judy Noe, who had the vision to see the true forest of the book amidst the tangle of trees.

But most of all, we would like to thank the thousands of participants in our workshops and classes for inviting us to share their personal and professional joys, frustrations, and insights.

CONTENTS

Acknowledgments   vi
Preface   xv

Part I: Negotiating a Common Understanding   1

1.   Methodologies Aren't Enough   3
    1.1 CASE, CAD, and the Cockroach Killer   3
    1.2 Methods' Effects on Problems   4
    1.3 Maps and Their Notation   6
    1.4 Making Sure That Everyone Can Read the Map   8
    1.5 Maps of Requirements Are Not Requirements   9
    1.6 Helpful Hints and Variations   9
    1.7 Summary   12

2.   Ambiguity in Stating Requirements   14
    2.1 Examples of Ambiguity   14
        2.1.1 *Missing requirements*   16
        2.1.2 *Ambiguous words*   16
        2.1.3 *Introduced elements*   16
    2.2 Cost of Ambiguity   17
    2.3 Exploring to Remove Ambiguity   19
        2.3.1 *A picture of requirements*   19
        2.3.2 *A model of exploration*   20
    2.4 Helpful Hints and Variations   21
    2.5 Summary   21

3.   Sources of Ambiguity   22
    3.1 An Example: The Convergent Design Processes Lecture   22
    3.2 A Test for Attentiveness   25
    3.3 The Clustering Heuristic   26
        3.3.1 *Observational and recall errors*   28
        3.3.2 *Interpretation errors*   28
        3.3.3 *Mixtures of sources of error*   29
        3.3.4 *Effects of human interaction*   29
    3.4 Problem Statement Ambiguity   30

3.5 Helpful Hints and Variations   32
3.6 Summary   32

4. The Tried but Untrue Use of Direct Questions   34
    4.1 Decision Trees   34
         4.1.1 *Order of questions*   36
         4.1.2 *Traversing the decision tree: an example*   36
    4.2 Results of an Ambiguity Poll   42
    4.3 What Could Possibly Be Wrong?   43
    4.4 Real Life Is More Real Than We Like to Think   44
    4.5 Helpful Hints and Variations   44
    4.6 Summary   45

Part II: Ways to Get Started   47

5. Starting Points   49
    5.1 A Universal Starting Point   49
    5.2 Universalizing a Variety of Starting Points   50
         5.2.1 *Solution idea*   50
         5.2.2 *Technology idea*   51
         5.2.3 *Simile*   52
         5.2.4 *Norm*   53
         5.2.5 *Mockup*   53
         5.2.6 *Name*   54
    5.3 The Can-Exist Assumption   55
    5.4 An Elevator Example   55
         5.4.1 *Naming our project*   56
    5.5 Helpful Hints and Variations   57
    5.6 Summary   58

6. Context-Free Questions   59
    6.1 Context-Free Process Questions   59
    6.2 Potential Impact of a Context-Free Question   60
    6.3 Context-Free Product Questions   61
    6.4 Metaquestions   62
    6.5 Advantages of Context-Free Questions   64
    6.6 Helpful Hints and Variations   65
    6.7 Summary   67

7. Getting the Right People Involved   68
    7.1 Identifying the Right People   68
         7.1.1 *Customers versus users*   68
         7.1.2 *Why include the users?*   69
         7.1.3 *The Railroad Paradox*   69
         7.1.4 *The product can create users*   70
         7.1.5 *Are losers users?*   71
    7.2 A User-Inclusion Heuristic   72
         7.2.1 *Listing possible user constituencies*   72

      7.2.2 Pruning the user list   73
  7.3 Participation   74
      7.3.1 Who participates?   74
      7.3.2 When do they participate?   76
      7.3.3 How do we get their judgments?   76
  7.4 Plan for Capturing Users   77
  7.5 Helpful Hints and Variations   78
  7.6 Summary   78

8.   Making Meetings Work for Everybody   80
  8.1 Meetings: Tools We Can't Live With, or Without   80
      8.1.1 A terrible, but typical, meeting   80
      8.1.2 Meetings as measurements   83
  8.2 Participation and Safety   84
      8.2.1 Establishing an interruption policy   84
      8.2.2 Setting time limits   84
      8.2.3 Outlawing personal attacks and put-downs   85
      8.2.4 Reducing pressure   85
      8.2.5 Allowing time to finish, yet finishing on time   86
      8.2.6 Handling related issues   86
      8.2.7 Amending the rules   86
  8.3 Making It Safe *Not* to Attend a Meeting   86
      8.3.1 Publishing an agenda and sticking to it   87
      8.3.2 Staying out of emergency mode   87
      8.3.3 Handling people who don't belong   87
      8.3.4 Including the right people   89
  8.4 Designing the Meeting You Need   89
  8.5 Helpful Hints and Variations   89
  8.6 Summary   91

9.   Reducing Ambiguity from Start to Finish   92
  9.1 Using the Memorization Heuristic   92
  9.2 Extending the Ambiguity Poll   93
  9.3 "Mary had a little lamb" Heuristic   94
  9.4 Developing the "Mary conned the trader" Heuristic   95
  9.5 Applying the Heuristics to the Star Problem   97
  9.6 Helpful Hints and Variations   101
  9.7 Summary   102

Part III: Exploring the Possibilities   105

10. Idea-Generation Meetings   109
  10.1 A Typical Brainblizzard   109
  10.2 First Part of the Brainstorm   111
      10.2.1 Do not allow criticism or debate   111
      10.2.2 Let your imagination soar   112
      10.2.3 Shoot for quantity   112
      10.2.4 Mutate and combine ideas   114

10.3 Second Part of the Brainstorm   115
 10.3.1 *Voting with a threshold*   116
 10.3.2 *Voting with campaign speeches*   117
 10.3.3 *Blending ideas*   117
 10.3.4 *Applying criteria*   117
 10.3.5 *Scoring or ranking systems*   117
10.4 Helpful Hints and Variations   117
10.5 Summary   118

11. Right-Brain Methods   120
11.1 Mapping Tools   120
 11.1.1 *Sketching*   120
 11.1.2 *Sketching Wiggle Charts*   122
11.2 Braindrawing   123
11.3 Right-Braining   123
11.4 Helpful Hints and Variations   125
11.5 Summary   127

12. The Project's Name   128
12.1 Working Titles, Nicknames, and Official Names   128
12.2 The Influence of Names   129
 12.2.1 *A naming demonstration*   129
 12.2.2 *What naming accomplishes*   131
12.3 The Naming Heuristic   132
12.4 Helpful Hints and Variations   134
12.5 Summary   134

13. Facilitating in the Face of Conflict   136
13.1 Handling Inessential Conflicts   136
 13.1.1 *Wrong time, wrong project*   137
 13.1.2 *Personality clashes*   137
 13.1.3 *Indispensable people*   138
 13.1.4 *Intergroup prejudice*   139
 13.1.5 *Level differences*   139
13.2 The Art of Being Fully Present   140
13.3 Handling Essential Conflicts   141
 13.3.1 *Reframing personality differences*   141
 13.3.2 *Negotiating*   143
 13.3.3 *Handling political conflicts*   144
13.4 Helpful Hints and Variations   144
13.5 Summary   145

Part IV: Clarifying Expectations   147

14. Functions   149
14.1 Defining a Function   149
 14.1.1 *Existence function*   149
 14.1.2 *Testing for a function*   150

14.2   Capturing All and Only Functions   150
     14.2.1   *Capturing all potential functions*   150
     14.2.2   *Understanding evident, hidden, and frill functions*   152
     14.2.3   *Identifying overlooked functions*   155
     14.2.4   *Avoiding implied solutions*   156
     14.2.5   *The ''Get It If You Can'' List*   157
14.3   Helpful Hints and Variations   158
14.4   Summary   159

15.   Attributes   161
15.1   Attribute Wish List   161
15.2   Transforming the Wish List   163
     15.2.1   *Distinguishing between attributes and attribute details*   163
     15.2.2   *Uncovering attribute ambiguity*   163
     15.2.3   *Organizing the list*   164
     15.2.4   *Discovering insights from the transformed list*   165
15.3   Assigning Attributes to Functions   166
     15.3.1   *How attributes can modify functions*   166
     15.3.2   *Gaining insights from the new format*   167
15.4   Excluding Attributes   167
     15.4.1   *Categorizing into must, want, and ignore attributes*   167
     15.4.2   *Implicit versus explicit elimination of attributes*   168
15.5   Helpful Hints and Variations   169
15.6   Summary   169

16.   Constraints   171
16.1   Defining Constraints   171
16.2   Thinking of Constraints as Boundaries   172
16.3   Testing the Constraints   174
     16.3.1   *Too strict?*   174
     16.3.2   *Not strict enough?*   175
     16.3.3   *Unclear?*   176
     16.3.4   *Generating new ideas*   176
16.4   Interrelated Constraints   177
16.5   Overconstraint   179
16.6   Psychology of Constraints   180
     16.6.1   *The tilt concept*   180
     16.6.2   *Breaking constraints*   181
     16.6.3   *The self-esteem bad-design cycle*   182
16.7   Constraint Produces Freedom   182
     16.7.1   *Standards*   182
     16.7.2   *Languages and other tools*   183
16.8   Helpful Hints and Variations   183
16.9   Summary   185

17.   Preferences   186
17.1   Defining a Preference   187
     17.1.1   *An example*   187

17.1.2   *The origin of preferences*   188
17.2   Making Preferences Measurable   188
17.2.1   *A reasonable approach to metrics*   188
17.2.2   *Making the preference measurable*   189
17.3   Distinguishing Between Constraints and Preferences   189
17.3.1   *Is meeting the schedule a constraint?*   190
17.4   Constrained Preferences   191
17.4.1   *What's-it-worth? graphs*   192
17.4.2   *When-do-you-need-it? graphs*   193
17.5   Reframing Constraints into Preferences   194
17.5.1   *Trading off among preferences*   195
17.5.2   *Zeroth Law of Product Development*   198
17.6   Helpful Hints and Variations   198
17.7   Summary   199

18.   Expectations   201
18.1   Reasons to Limit Expectations   201
18.1.1   *People are not perfect*   202
18.1.2   *People are not logical*   202
18.1.3   *People perceive things differently*   202
18.1.4   *Designers are people, too*   204
18.2   Applying the Expectation Limitation Process   204
18.2.1   *Generate a specific expectation list*   204
18.2.2   *The elevator example*   204
18.2.3   *Generalize the expectation list*   206
18.2.4   *Limit the expectations*   208
18.3   Limitations Need Not Be Limiting   209
18.3.1   *The wheelchair example*   209
18.3.2   *Keeping possibilities open*   209
18.3.3   *Include the source of the limitation*   210
18.4   Helpful Hints and Variations   210
18.5   Summary   212

Part V: Greatly Improving the Odds of Success   215

19.   Ambiguity Metrics   217
19.1   Measuring Ambiguity   217
19.1.1   *Using the ambiguity poll*   217
19.1.2   *Applying the polling method*   218
19.1.3   *Polling on different bases*   218
19.2   Using the Metric as a Test   219
19.2.1   *Measuring three kinds of ambiguity*   219
19.2.2   *Interpreting the results*   220
19.2.3   *Information from clustering*   221
19.2.4   *Choosing the group to be polled*   221
19.3   Helpful Hints and Variations   222
19.4   Summary   223

20. Technical Reviews   225
     20.1 A Walkover Example   225
     20.2 The Role of Technical Reviews   227
           20.2.1 Formal and informal reviews   227
           20.2.2 Technical reviews versus project reviews   228
     20.3 Review Reports   230
           20.3.1 Need for review reports   230
           20.3.2 Creating the issues list   231
           20.3.3 Technical review summary report   232
     20.4 Principal Types of Review   232
           20.4.1 Vanilla reviews   233
           20.4.2 Inspections   234
           20.4.3 Walkthroughs   234
           20.4.4 Round robin reviews   234
     20.5 Real Versus Ideal Reviews   235
     20.6 Helpful Hints and Variations   236
     20.7 Summary   236

21. Measuring Satisfaction   238
     21.1 Creating a User Satisfaction Test   238
           21.1.1 Test attributes   238
           21.1.2 A custom test for each project   239
     21.2 Using the Test   240
           21.2.1 Benefits   241
           21.2.2 Plotting shifts and trends   241
           21.2.3 Interpreting the comments   242
           21.2.4 Feelings are facts   242
     21.3 Other Uses of the Test   244
           21.3.1 A communication vehicle   244
           21.3.2 Continued use throughout the project   245
           21.3.3 Use by designers   245
     21.4 Other Tests   246
           21.4.1 Prototypes as satisfaction tests   246
     21.5 Helpful Hints and Variations   247
     21.6 Summary   248

22. Test Cases   249
     22.1 Black Box Testing   249
           22.1.1 External versus internal knowledge   249
           22.1.2 Constructing black box test cases   250
           22.1.3 Testing the test cases   251
     22.2 Applying the Test Cases   252
           22.2.1 Examples   252
           22.2.2 Iterating tests and answers   254
           22.2.3 Clearly specifying ambiguity   255
     22.3 Documenting the Test Cases   256
     22.4 Helpful Hints and Variations   257
     22.5 Summary   258

23. Studying Existing Products   260
     23.1 Use of the Existing Product as the Norm   260
     23.2 Interviewing   261
           23.2.1 *What's missing in the new product?*   262
           23.2.2 *Why is it missing?*   262
           23.2.3 *What's missing in the old product?*   263
           23.2.4 *What's missing in the old requirements?*   263
     23.3 Substituting Features for Functions   264
     23.4 Helpful Hints and Variations   266
     23.5 Summary   266

24. Making Agreements   268
     24.1 Where Decisions Come From   268
           24.1.1 *Choices, assumptions, and impositions*   268
           24.1.2 *Elevator design decision examples*   269
           24.1.3 *Writing traceable requirements*   271
     24.2 Where False Assumptions Come From   271
           24.2.1 *Lack of valid information*   272
           24.2.2 *Invalidation over time*   272
           24.2.3 *The turnpike effect*   272
           24.2.4 *Requirements leakage*   272
     24.3 Converting Decisions to Agreements   273
     24.4 Helpful Hints and Variations   274
     24.5 Summary   275

25. Ending   277
     25.1 The Fear of Ending   277
     25.2 The Courage to End It All   277
           25.2.1 *Automatic design and development*   278
           25.2.2 *Hacking*   279
           25.2.3 *Freezing requirements*   280
           25.2.4 *The renegotiation process*   280
           25.2.5 *The fear of making assumptions explicit*   281
     25.3 The Courage to Be Inadequate   282
     25.4 Helpful Hints and Variations   283
     25.5 Summary   283

Bibliography   285
Index   293

PREFACE

"There's no sense being exact about something if you don't even know what you're talking about."

—John von Neumann

Development is the process of transforming someone's desires into a product that satisfies those desires. This book is about the *requirements process*—the part of development in which *people attempt to discover what is desired.*

To understand this process, the reader should focus on five critical words: desire, product, people, attempt, and discover.

First, consider the word "desire." Some readers would prefer that we say "attempt to discover what is *needed*," but we don't know how to figure out what people *need*, as opposed to what they *desire*. Besides, people don't always buy what they *need*, but they always *desire* what they buy, even if the desire is only transitory. We do observe, however, that by clarifying their desires, people sometimes clarify what they really need and don't need.

By "product," we mean any product that attempts to satisfy a complex set of desires. One reason the desires are complex is that they come from many people. When we create a product to satisfy our own desires—as when we build a garden, for example, or a bookcase—we don't often need an explicit requirements process. We simply build something, look at it, and make adjustments until we are satisfied.

But "people" might include many different people, and discovering who "people" are is a major part of the requirements process. When many people are involved—and when the product is large—the kind of iterative process used to discover personal requirements may simply prove too drawn out, too costly, and too risky.

What about "attempt"? If we're writing a book, shouldn't we be more certain, more positive? Shouldn't we *guarantee* results? Well, we've used the requirements techniques in this book to help our clients develop all types of products—computer hardware, computer software, automobiles, furniture, buildings, innovative consumer products, books, films, organizations, training courses, and research plans. Nobody yet has demanded money back, but we can't prove no client ever will, because we do not know how to make product development into an exact discipline.

Before working with us, many of our clients hoped that product development was an exact discipline. Many of those clients were in the software business—a business that has been plagued by ill-founded fantasies of an exact discipline for developing products. We like to quote John von Neumann because many of our clients consider him to be the founding father of software. They pay attention when he says, *"There's no sense being exact about something if you don't even know what you're talking about."*

If people don't know what they want, no development process—no matter how exact, how clever, or how efficient—will satisfy them. And that's why we do requirements work—*so we don't design systems that people don't want.*

Effectiveness always comes before *efficiency.* But even if efficiency is your hot button, the most important route to efficiency in development is to eliminate those products nobody wanted in the first place. Another way to put this is,

Anything not worth doing is not worth doing right.

Which brings us to "discover," the most critical word. In this book, we're trying to help people discover what's really worth doing.

Dwight Eisenhower once said, "The plan is nothing; the planning is everything." We agree, and we also extend the same thinking to the requirements process:

The product is nothing; the process is everything.

Or, put another way,

The discovery is nothing; the discovering (the exploring) is everything.

which explains the title, *Exploring Requirements.*

A data dictionary, for example, is a way of preserving the definitions that are painstakingly derived with the aid of some of the methods in this book. In practice, however, hardly anybody reads the data dictionary, or possibly any of the documents that are developed in the requirements process. That observation bothers some people, but not us because we believe that

The document is nothing; the documenting is everything.

If you watch how people *really* develop systems, you'll see that the process of developing requirements is actually a process of developing a team of people who

1.   understand the requirements

2.   (mostly) stay together to work on the project

3.   know how to work effectively as a team

We believe that if one of these conditions is not met, the project will probably fail. Of course, there are many other reasons why a product development project might fail, and there are many books about methods to avoid such pitfalls. This book, however, will concentrate on these three critical but neglected human aspects of the requirements process:

1. developing a consistent understanding of requirements among all participants

2. developing the desire to work as a team on the project

3. developing the necessary skills and tools for working effectively as a team to define requirements

Because these topics are somewhat neglected in the systems development literature, *Exploring Requirements* can be used as a supplement to any requirements process that you now use, formal or informal. Most of the chapters are designed as stand-alone modules, each presenting one or more tools or methods to enhance your requirements process. You can read the book from cover to cover, or read only the one chapter that's most needed at any given moment. Either way, it should help you do a better job of knowing what you're talking about.

EXPLORING REQUIREMENTS: QUALITY BEFORE DESIGN

PART I   NEGOTIATING A COMMON UNDERSTANDING

Before you invest your time in a book, you want to know what you'll get from reading it. This book is about development projects. In particular, it's mostly about the *early* part of development projects, or at least what should be the early part.

Similarly, in the early part of a development project before you are willing to invest too much of your time, you want to know what you'll get from doing it. Well, unless you know what you *want*, you'll never be very sure of what you'll *get*.

If you employ other people to help you develop what you want, you'd describe what you want for them. That description is called a *problem statement* or a *set of requirements*, and that's what this book is about. Obviously, requirements are important because if you don't know what you want, or don't communicate what you want, you reduce your chances of getting what you want.

Some years ago, we performed experiments to determine how computer programmers were influenced by what they were asked to do.* The experiments were simple, but they revealed the essential dynamics underlying all requirements work:

- If you tell what you want, you're quite likely to get it.

- If you don't tell what you want, you're quite unlikely to get it.

In a typical experiment, five teams were given the same requirement for a computer program except for a single sentence that differed for each team. One team was asked to complete the job with the fewest possible hours of programming, another was to minimize the number of program statements written, a third was to minimize the amount of memory used, another was to produce the clearest possible program, and the final team was to produce the clearest possible output. The actual results are shown in Figure I-1.

---

*G.M. Weinberg and E.L. Schulman, "Goals and Performance in Computer Programming," *Human Factors*, Vol. 16, No. 1 (1974), pp. 70-77. Reprinted in Bill Curtis, ed. *Tutorial: Human Factors in Software Development* (Los Angeles: IEEE Computer Society, 1981).

Primary Objective	Rank on Primary
minimize core storage	1
maximize output readability	1
maximize program readability	1-2
minimize statements	1
minimize programming hours	1

Figure I-1. Experiments show that programming team performance is highly sensitive to stated requirements.

In short, each team produced *exactly what it was asked to produce,* and didn't produce what it wasn't asked to produce. Before these experiments, we had often heard software buyers complaining about the inability of programmers to give them what they wanted. The experiments convinced us that in many cases, the buyers simply *did not tell the programmers clearly what they wanted.*

In the ensuing years, we have confirmed this observation in software development organizations all over the world. Moreover, we've found that the difficulty in stating requirements is not confined to software development, but is found whenever people design and build products for other people. Between your two authors, we've devoted more than sixty years to overcoming this difficulty. We've developed techniques to help people negotiate a common understanding of what they want. If you require such techniques, then this book was designed for you.

1 METHODOLOGIES AREN'T ENOUGH

Some of our readers would readily agree with the need to resolve the ambiguities in requirements, but they would argue that the problem is not with the techniques, but with the *technicians*. In order to do a better requirements job, they would claim, remove the people from the process and instead use a methodology. Indeed, their preference would be a totally *automated* methodology, with no people in the development process whatsoever.

When we started teaching software development in 1958, there were few organized development methods. Today, however, packaged methodologies flood the market and almost everyone who develops software has some sort of organized process. Now, computer-aided software engineering (CASE) and computer-aided design (CAD) dominate the news, both promising to eliminate people from the process. So why do we still need people to explore requirements? Who needs a book about "people-oriented" tools? Won't CASE and CAD tools ensure that everyone gets the right requirements, and gets the requirements right? We think not, and the story of the Guaranteed Cockroach Killer will tell you why.

1.1 CASE, CAD, and the Cockroach Killer

For many years, a man in New York made a living selling his Guaranteed Cockroach Killer through the classified ads. When you sent your check for five dollars, he cashed it and sent you the kit shown in Figure 1-1.

The instructions read

1. Place cockroach on block A.

2. Hit cockroach with block B.

If you follow these instructions, you are guaranteed to kill the cockroach. By rights, there should have been a third instruction:

3. Clean up mess.

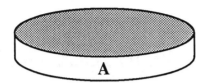 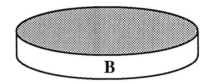

Figure 1-1. The Guaranteed Cockroach Killer kit. All you have to do is place the roach on block A and then strike it with block B.

Quite likely, nobody ever needed the third instruction—which also meant that nobody could collect on the iron-clad guarantee.

Now, what has the Guaranteed Cockroach Killer to do with CASE and CAD tools? In one CASE document, we read that CASE tools are comprised of three basic application development technologies:

1.   analysis and design workstations that facilitate the capture of specifications

2.   dictionaries that manage and control the specification information

3.   generators that transform specifications into code and documentation

In our experience, "analysis and design workstations" resemble block A of the cockroach killer. "Dictionaries" resemble block B. If you can somehow figure out the specifications, the workstations will "capture" them and the dictionaries will "manage" them. Once these two steps are done, the generators will clean up the mess and provide your product.

These CASE and CAD tools are guaranteed to do *their* job if you simply do your part. This book is about doing *your* part: getting the roach on the block, and convincing it to stand still long enough for you to deliver the crushing blow.

Whether it's roaches or requirements, the hard part is catching them and getting them to stand still. Generating the code and cleaning up the squashed roach are *messy* jobs, but in a different way than the first two steps. That's why we think that even with *guaranteed* CASE and CAD tools, you'll still need the tools described in this book. Not only that, but the better your CASE tools, the more you'll need our tools. Let's see why.

1.2 Methods' Effects on Problems

Without a doubt, formal methods—especially automated formal methods—have transformed systems development work. Today, we routinely watch our clients build systems in hours that would have taken weeks to build in 1958. Yet even though projects take a fraction of the time, developers still find themselves wallowing for years in other projects. Because there are now superior tools and methods, developers attempt systems that never could have been imagined in 1958.

Not only that, but within the same systems the problems that are most amenable to formal solutions are quickly eliminated, leaving a residue of "messy" problems that are not amenable to formal solutions. As a result of these two effects, the nature of the problem has been transformed, as suggested in Figure 1-2.

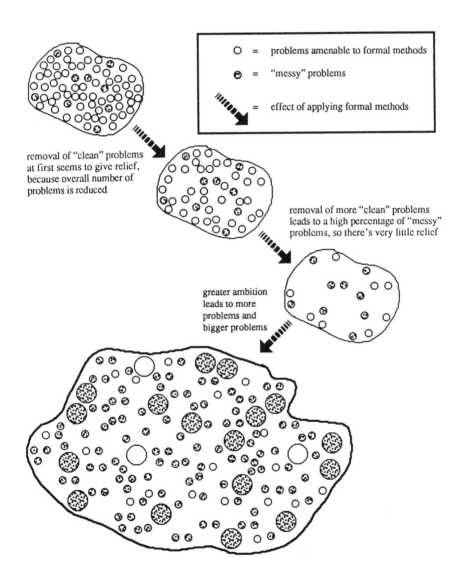

Figure 1-2. How methods change problems: First, problems amenable to formal methods are eliminated, leaving a higher percentage of "messy" problems. Then, because of the promise of the formal methods, ambition grows, leaving more problems of every kind.

People who have worked in product and systems development for many years have noticed this transformation in their own daily work. They spend less time on technical tasks and more time dealing with people, less time on small closed tasks and more time working on large ones that never seem to finish. Our work, as well as this book, is devoted to tackling those "messy" problems that the formal methods don't touch, as well as the more ambitious problems that arise from expecting too much from the formal methods.

1.3 Maps and Their Notation

Our methods have been applied in conjunction with many CASE and CAD tools and many different formal methods of development. We've worked with information systems approaches such as Warnier-Orr, Jackson, Ross's SADT™, Yourdon-Constantine, and Gane-Sarson; but we've also used our tools with more traditional engineering disciplines such as electrical engineering, mechanical engineering, architecture, and city planning.

An important part of any method of developing systems is the *notation*, its own symbolic way of representing system ideas. In terms of our exploration metaphor, we would say there are many systems of map-making. In city planning, many of the maps are recognizable as conventional maps. In architecture, some of the maps, such as floor plans and sketches, are quite literal pictures of what they represent. Others, such as wiring diagrams and materials lists, are rather abstract. Renderings of machines and parts are maps, and so are tables of strengths of materials.

Figures 1-3 and 1-4 are two different maps of the same process—writing and sending a letter. Figure 1-3 uses pictures, and is easier to understand at one glance. The words in Figure 1-4 prevent a quick and universal understanding, but also avoid perhaps unintended implications that the pictures may convey.

For instance, the typewriter in Figure 1-3 is a manual typewriter, whereas Figure 1-4 leaves the nature of the machine open. If it were important that an electric typewriter be used, the notation of Figure 1-4 would allow us to write, "electric typewriter." In Figure 1-3, we would have to draw a different picture. Which notation is better? Although CASE proponents argue interminably, there is no one answer to this question because each notation shows the system in different ways. Therefore, both views can be useful, and even necessary.

Another important property of a good notation is that diagrams can be revised easily, so as not to interfere with the fluidity of the requirements process. We wouldn't want to leave an overly rigid implication in the map just because it would be too much bother to change it. This is when a good CASE and CAD tool can shine: by allowing us to keep the most current maps in front of us all of the time.

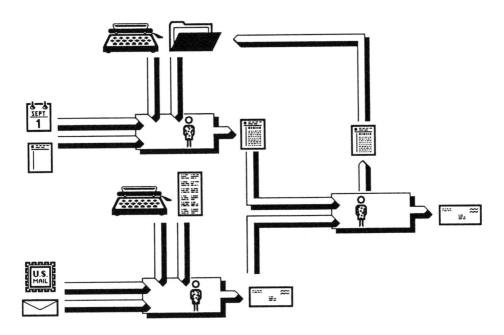

Figure 1-3. A stylized map of the process of writing a letter. The graphic form of this map makes it easier to apprehend at a glance during a presentation, but may make it difficult to modify as changes are made.

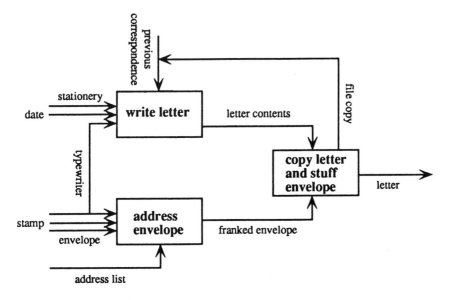

Figure 1-4. Another map of writing a letter, in a notation that is less visual than Figure 1-3, but more amenable to modification.

1.4 Making Sure That Everyone Can Read the Map

The particular symbols used in mapping systems are not as important as their proponents would have you believe. Each notation has advantages, disadvantages, and a variety of tools for working with it, but the most important tool to work with a map is the human mind. That's why *the most important quality of a map is that everyone involved should be able to understand it.*

Proponents of each notation claim that their maps are "intuitive," and "easy to read." These statements are true in the same sense that Chinese is intuitive—in Beijing. Virtually any notational system becomes intuitive after someone has spent a lot of time working with it. In requirements work, however, most of the participants will not be professional requirements writers, and they may be working with such maps for the first time.

In light of the "amateur" standing of most participants, at least some of the maps should be familiar to everyone without special training. When special notations must be used, plan for a familiarization period. There are several exercises for this purpose.

1. Have each map presented by someone who does *not* know the notation. This approach reveals ambiguities in the notation, but also reveals ambiguities in the specification itself.

2. Because each method shows and conceals different aspects, ideally everyone should know several methods of mapping. To accomplish this, ask each person to translate a map from one notation to another. In translating Figure 1-3 to Figure 1-4, the translator assumed that the same typewriter was used for writing the letter and addressing the envelope. This assumption prompted a discussion, which revealed other assumptions about how the letter and envelope would be prepared.

3. Even better would be to use a CASE or CAD system with several embedded notations and the ability to transform a map from one to the other, instantly. With such a system, the same maps could be presented in different ways so that each participant could see a familiar form. On the other hand, this completely automatic process does not really substitute for methods 1 and 2, which force more intimate involvement with the map and thus foster deeper learning.

4. The most common method for addressing the problem of familiarization (when it is addressed at all) is to delay the start of requirements work until everyone has attended a course on the common notation. Although we applaud the recognition of this problem,

we deplore the solution method. By the time everyone is scheduled, and then rescheduled because of changed plans, your requirements process will have lost whatever momentum it had. Better to conduct the class as part of the requirements meetings, perhaps keyed to specific examples from the system being developed.

1.5 Maps of Requirements Are Not Requirements

In exploring requirements, developers work with *maps* of requirements, not the requirements themselves. While working with a client in Sweden, we learned a useful rule that's taught to every recruit in the Swedish army:

**When the map and the territory don't agree,
always believe the territory.**

When participants get wrapped up in formal systems development, especially when supported by automatic mapping tools, they sometimes forget the territory and begin to believe that the map is the territory. For example, in using Figure 1-3, you might unconsciously assume that the person doing the steps is a woman because the symbolic figure seems to be a female. You might also assume that only one person works on a step, and that it is the *same* woman in each step, because the same symbol is used on the map.

Or, using Figure 1-4, you might assume that the file copy must be developed in the same step as stuffing the envelope, which would preclude a system in which writing the letter produced the copy—as with carbon paper or multicopy laser printers. Figure 1-5 is another diagram in the same notation, showing a process in which the copy is produced in the same step as the letter-writing process.

Figure 1-6 shows a version of Figure 1-5 in another popular notation for describing processes. Some people believe that this notation is clearer than Figure 1-5, others believe the reverse. We believe that it's mostly a matter of taste, and experience with each notation. But most of all, it's a matter of *purpose*—what types of exploration are being done, and what types of errors are being targeted to avoid.

1.6 Helpful Hints and Variations

• Another source of complexity in development projects is the desire of nonexperts to be involved in the design of complex products. Most users naturally prefer to learn as little as possible about product design, yet still get all their wishes met in the requirements process. When users prefer to remain naive about the product design process, most of the burden falls on the notation, and that's a lot to ask of a notation. The design complexity problem can be solved by creating a tutorial to convert a naive user into an expert—but not if the user doesn't want to invest that kind of effort.

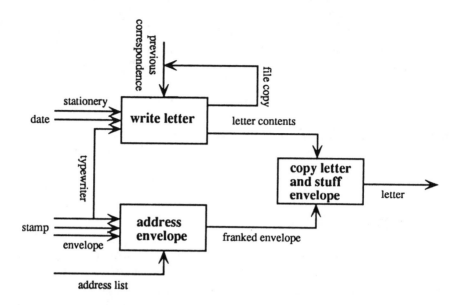

Figure 1-5. Yet another map of writing a letter, in the notation of Figure 1-4, but with a different handling of the filing process. Both these maps may show more precision than was intended, as no specific manner of filing may have been implied.

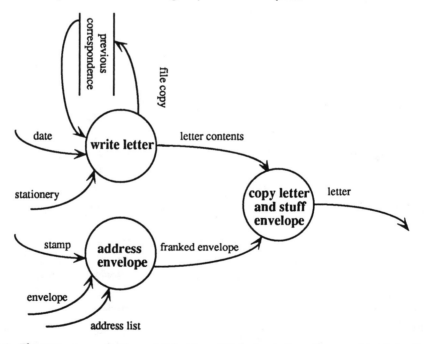

Figure 1-6. The same map as in Figure 1-5, but in a different notation. Here, nothing is implied about the devices used to accomplish the actions, such as the typewriter. Some people prefer the curved lines, and some the straight, but is it worth destroying a project over?

- One reason customers turn away from the design process is that the professional designers treat them in a patronizing manner. Remember that most participants are naive only about the development process, and that they are experts themselves in subjects about which the designers are naive. That's why their participation is needed. We're much more likely to get that participation if we create an environment of shared expertise.

- Methodology experts underestimate the difficulty of understanding their notations, which they believe to be "intuitive." To understand how difficult it really is to define an "intuitive" notation, consider the problem of designing international road signs that are universally understood.

- When two sets of experts participate in the same requirements process, there's often a conflict about which notation is intuitive. Children reared in Paris think that French is intuitive, but children reared in Montreal may think that both French and English are intuitive. In the same way, experts often share a "first love" syndrome for notations.

  This bias can be prevented in the same way that bilingual children avoid a bias for one language. If you are teaching people a notation, instruct them in two at the same time. Do all maps in both notations, and compare the pair of maps explicitly.

- Mastery of the notation may give one party dominance in the requirements process, and exclude those who haven't mastered the notation. If you want to avoid political wars, you must devote attention to depoliticizing the language issue. Try following the Swiss example: All notations that are "first love" notations of participants are declared to be "official" languages of the requirements process. Every official document must be presented in all of the official languages before the process is considered complete.

  Although this multilingual approach may seem burdensome, the Swiss example shows that it can work if done in the right spirit. Indeed, in Swiss meetings, people from the "dominant" language group often present their ideas in the "secondary" languages, as a courtesy to the minority participants and as a strong indication of how much they value their participation. In our experience with requirements work, this multilingual approach *always* yields a more accurate definition of requirements as well as a more complete involvement of all participants.

- Our colleague Eric Minch suggests a theoretical model in which *all* descriptions of the designed system—including the various constituencies' requirements and satisfactions, the constraints and decisions, and the final specifications for implementation—can be considered statements in different "languages." In other words, instead of seeing all of these as different descriptions in a single language, we can think of them as the same description in different languages. The full design task then involves finding a way of translating between these languages, and the final product is such a translation strategy.

This idea stimulated us to observe that "translation" is not quite the simple one-to-one task that we often assume. Some translations are works of art in themselves, like Edward FitzGerald's English translation of *The Rubáiyát of Omar Khayyám*. Consider the famous quatrain:

> A Book of Verses underneath the Bough,
> A Jug of Wine, a Loaf of Bread—and Thou
>   Beside me singing in the Wilderness—
> Oh, Wilderness were Paradise enow!

What part is original with Omar the Tentmaker, astronomer and mathematician from eleventh century Persia? What part has been added (or subtracted) by FitzGerald, the nineteenth century Suffolk gentleman? What part has been added by us, in reading?

Experiencing the final product, we don't so much care if it's an exact translation, but only that we like the product. This is a reminder that value can be added at *any* stage of the product development cycle, and that requirements are only a guide. They are to be taken literally, but not too literally. There are many roads to Paradise.

- These observations connect directly with a remark by Ken de Lavigne, our colleague at IBM: "Your discussion of maps brings out the great problem introduced by 'stepwise refinement' approaches: although they claim to postpone decisions until the time comes to make them, in fact the most far-reaching decisions are made first, when one has the least amount of information."

Always be prepared to go back and revise the "translation," or "map," when further exploration shows that there was a wrong branch higher in the decision tree. To do this, let go of the idea of the "one right way," or "the perfect translation."

1.7 Summary

Why?
"Exploring" comes from the idea of working with maps of requirements, not the requirements themselves. People explore to make maps, and eventually get a map that is close enough to the territory to represent it for "practical" purposes.

When?
Always work with maps, not with the territory itself, and so *always* work to ensure common understanding of the various maps used. The question, "Could you please explain your notation?" is always in order in the requirements process, and nobody should be made to feel foolish or uncooperative for asking it.

How?

Mapping is not so much a process as it is an awareness and a commitment to clear communication. To avoid communication problems, consider the following:

1.   Don't leave all the hard parts to be handled by someone else, as in the Guaranteed Cockroach Killer kit.

2.   Be prepared for changes that new methods will introduce, and especially don't underestimate the difficulties for some people in adapting to new methods and new notations.

3.   Recognize that people are different, and that each mapping technique will appeal in different ways to every participant. Accept these differences, and don't try to force people to conform or belittle them for being "stupid."

4.   Instead, make sure at every step that everyone can read the map in order not to exclude anyone.

5.   If you find yourself getting touchy about notational nuances, remember that the map is not the territory and that there's no such thing as a perfect translation.

Who?

Everybody uses maps, though not everybody is equally facile with each system of mapping. Those who are more facile have the responsibility to assist those who are less so, and not in a condescending way.

2 Ambiguity in Stating Requirements

Whenever you use tools that ignore the human aspect, you describe requirements imperfectly and create ambiguities. Ambiguities, in turn, lead to diverse interpretations of the same requirement.

2.1 Examples of Ambiguity

For instance, Figures 2-1, 2-2, and 2-3 show three rather different structures built in response to the same ambiguous problem statement:

Create a means for protecting a small group of human beings from the hostile elements of their environment.

Figure 2-1. Igloo—an indigenous home constructed of local building materials.

Figure 2-2. Bavarian castle—a home constructed to impress the neighbors.

Figure 2-3. Space station—a mobile home with a view.

First of all, these three structures do represent effective solutions to the problem as stated, yet the solutions are strikingly different. Examining the differences among the solutions, we find clues to some of the ambiguity in the requirements.

2.1.1 Missing requirements

Sometimes requirements are missing. For instance, there is no requirement concerning *properties of materials,* properties such as local availability, durability, or cost. Thus, it's not surprising the three solutions differ widely in their use of materials.

The problem statement is equally ambiguous as to the *structure,* or how the building materials will be assembled. We don't know the desired size, shape, weight, or longevity of the structure.

Little is said or even implied about what *functions* will be performed inside these structures, leaving open the question of specific functional elements such as stoves, servants' quarters, beds, and ballrooms.

Nothing is said about the *physical environment,* either internal or external. The structure could reside on land, sea, or in the air, on an ice pack, or even in outer space. Then, too, we know nothing about the specific hostilities from which we are to protect the inhabitants.

What about the social and cultural environment? Is this small group of human beings a family unit, and if so, just what constitutes a family unit in this particular culture? Perhaps it is a working group, such as hunters or petroleum engineers, or possibly a recreational group, such as poker players or square dancers.

2.1.2 Ambiguous words

Even when requirements are stated explicitly, they may use ambiguous words. For instance, the word "small" does not adequately specify the *size of the group.* Beware of comparative words, like "small" or "inexpensive," in problem statements. A group of 25,000 would be "small" if we're talking about football fans at a University of Nebraska home game, where a new domed stadium could fulfill the stated requirements.

Another dangerous word in the problem statement is "group," which implies that the people will interact, somehow, but it's not clear *how.* A group of people performing *The Marriage of Figaro* don't interact in the same way as a group of people having coffee in the Figaro Café. Designing a structure for one group would be quite different from designing for the other.

Even the term "structure" carries a load of ambiguity. Some readers would infer that "structure" means something hard, durable, solid, opaque, and possibly heavy. If we slip unconsciously into that inference, we subliminally conclude the problem is to be solved with traditional building materials, thus limiting the range of possible effective designs.

2.1.3 Introduced elements

Of course, we can guard against ambiguous words by carefully exploring alternative meanings for each word in the problem statement, but that won't protect us from another problem. The term "structure" never actually appeared in the problem statement, but somehow slipped into our discussion without our noticing. The problem statement actually says "create a means," not "design a structure."

Some problem ambiguities are so obvious that they would be resolved in casual designer-client conversations long before the actual design process began. More subtle ambiguities, however, may be resolved unconsciously in the designer's mind. In this case, an innovative, but nonstructural, "means" of protecting a small group might be overlooked. For instance, the designer might

1. protect against rain by electrostatically charging the raindrops and repelling them with electrical fields.

2. protect against belligerent crowds by supplying aphorisms such as "Sticks and stones may break my bones, but names will never hurt me," or "I may have to respond to what other people do, but they don't define me."

3. protect against winter by moving south, and against summer by moving north.

2.2 Cost of Ambiguity

These few elementary examples of ambiguity in requirements can only suggest the enormous impact of the problem. Billions of dollars are squandered each year building products that don't meet requirements, mostly because the requirements were never clearly understood.

For instance, Boehm* analyzed sixty-three software development projects in corporations such as IBM, GTE, and TRW and determined the ranges in cost for errors created by false assumptions in the requirements phase but not detected until later phases. (See Table 2.1 and Figure 2-4.)

Table 2.1
Relative Cost to Fix an Error

Phase in Which Found	Cost Ratio
Requirements	1
Design	3-6
Coding	10
Development testing	15-40
Acceptance testing	30-70
Operation	40-1000

Although Table 2.1 vividly shows the importance of detecting ambiguities in requirements, the figures may actually be quite conservative. First of all, Boehm studied only projects that were completed, but some observers have estimated that ap-

---

*Barry W. Boehm, *Software Engineering Economics* (Englewood Cliffs, N.J.: Prentice-Hall, 1981).

proximately one-third of large software projects are never completed. Much of the enormous loss from these aborted projects can be attributed to poor requirements definition.

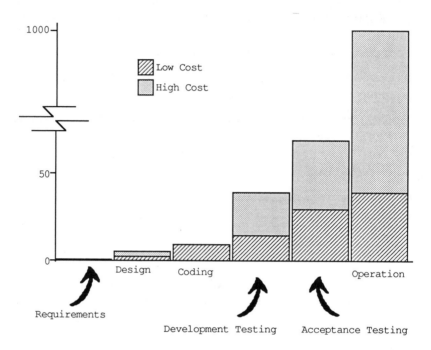

Figure 2-4. Boehm's observations on project cost.

The situation looks even worse when we consider the catastrophes resulting from incorrect design decisions based on early false assumptions. For example, on the Ford Pinto manufactured in the 1970s, the position of the fuel tank mounting bolts was a perfectly fine design decision based on an assumption, *there will be no rear impact collisions.* As this assumption proved to be false, the Ford Motor Company, by its own estimates, spent $100 million in litigation and recall services. And what value are we to place on the lost lives?

Or take another case. The decision by the Johns-Manville Corporation to develop, manufacture, and market asbestos building materials was based on the assumption that asbestos, in the form used in their products, was environmentally safe to all exposed people. Many fine ideas found their way into Johns-Manville products based on what we now understand to be a false assumption. Epidemiology Resources, Inc. of Cambridge, Massachusetts estimated that this high-level decision would eventually produce 52,000 lawsuits costing approximately $2 billion in liabilities. Indeed, the company's entire organization, culture, and capital investment was so dedicated to the production of asbestos materials that it went bankrupt and reorganized as the Manville Company.

2.3 Exploring to Remove Ambiguity

For the past three decades, we have both been working to help people avoid costly errors, failed projects, and catastrophes, many of which have arisen from ambiguous requirements. We have written this book to show you successful methods for *exploring requirements* in order to constrain ambiguity. These methods are general and can be applied to almost any kind of design project, whether it be a house in Peoria or a castle in Bavaria, an on-line information system or a manufacturing organization, an advertising campaign or a biking vacation in New Zealand.

2.3.1. A picture of requirements

Figure 2-5 is a picture illustrating what we mean by requirements. Many ancient peoples believed that the universe emerged from a large egg, so we've used the "egg" to represent the universe of everything that's possible. We've drawn a line at the middle of the egg to divide what we want from what we don't want. If we could actually draw, or describe, such a line, we would have a perfect statement of requirements.

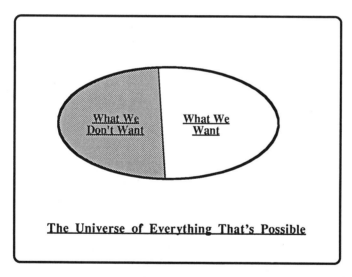

Figure 2-5. What we mean by requirements.

Figure 2-6 is a picture illustrating what we mean by *exploring* requirements. The first egg shows a rather wavy boundary that represents the first approximation to the requirements line. This line might represent the first vague statement of the problem. The second egg shows the results of some early exploration techniques. The line is still wavy, indicating there are still some things described that we don't want, and some not described that we do want. But at least some of the biggest potential mistakes (areas furthest from the requirements line) have been eliminated.

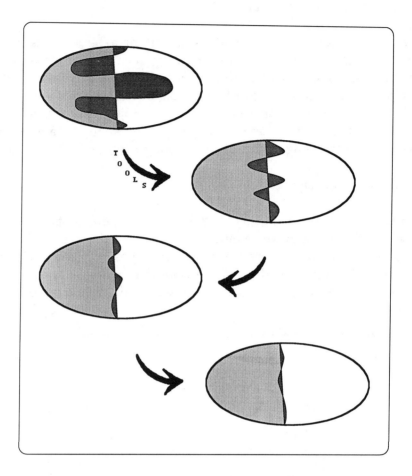

Figure 2-6. Exploring requirements: The black area represents what we want that we don't ask for, or what we ask for that we don't want. Through the repeated use of requirements tools, we reduce these areas and get closer to what we want, and only what we want.

Each new egg, which represents the next stage in the requirements process, produces a better approximation to the true requirements line. Unfortunately, the line will never match the true requirements perfectly, because in real life that's an almost impossible task. Through explorations, though, developers try to get it reasonably straight before paying too dearly for the wiggles.

2.3.2 A model of exploration

The process for straightening the wiggly line is an exploration, patterned after the great explorers like Marco Polo, Columbus, or Lewis and Clark. Roughly, here's what all explorers do:

1.   Move in some direction.

2.   Look at what they find there.

3.   Record what they find.

4.   Analyze their findings in terms of where they want to be.

5.   Use their analysis and recordings of what they find to choose the next direction.

6.   Go back to step 1 and continue exploring.

This is the same process used in exploring requirements, as we'll show in the rest of the book.

2.4 Helpful Hints and Variations

- Our colleague Ken de Lavigne suggests that some great examples of the consequences of requirements ambiguity are described in Deming's book, *Out of the Crisis*,* under the subject of "operational definition."

- Even better than looking at other people's examples is finding a few of your own. Next time you notice that a product isn't quite what you like, ask yourself, "What was the requirement that produced this way of creating the product?"

2.5 Summary

Why?
Attack ambiguity because ambiguity costs!

When?
Attack ambiguity as early as possible, because although you may eventually get rid of it, the cost of correction in early stages of product development is much, much less than later.

How?
How to attack ambiguity is the subject of the entire book. But above all, remember to use your mind in a playful way—exploration should be fun!

Who?
You, of course.

---

*W. Edwards Deming, *Out of the Crisis* (Cambridge, Mass.: MIT Center for Advanced Engineering Study, 1986).

3   SOURCES OF AMBIGUITY

Not all ambiguity comes from the same place, so once you recognize ambiguity in requirements, you'll still need to locate the source. To help you develop some ways of classifying the sources of ambiguity, let's pause to "participate" in one of our experiments in ambiguity. To get the most of this experiment, imagine that you are actually participating in the scene described.

3.1 An Example: The Convergent Design Processes Lecture

You've just driven twenty-three miles through heavy traffic to attend a professional seminar entitled "Convergent Design Processes, or Getting the Ambiguity Out," given by a Professor Donald C. Gause. You expect to spend a pleasant day sharing design war stories with the other participants, but otherwise you have little idea of what to expect. You don't have high regard for the practical sense of professors, but with luck, the presenter might not be too dull. Maybe you'll even get a few new ideas. After all, the seminar description promised help in identifying and defining the *real* design problem to be solved.

Your design consciousness is in high gear. You notice that it is 8:55 A.M. on your Ralph Lauren watch (designed to maximize snob appeal), and you have just settled comfortably into your stackable plastic chair (designed to maximize storage efficiency for the hotel). You warm your hands on a styrofoam coffee cup (designed to minimize cost per serving to the caterer) filled to the rim with something resembling coffee (designed to pacify late-rising coffee addicts). You're about to glance at your watch again when a distinguished-looking woman in a navy blue polyester suit and red bow tie strides to the lectern and makes a few remarks about restrooms, fire exits, refreshment breaks, and lunch. She then glowingly introduces Professor Gause, but you decide you'll wait to see for yourself.

The first thing you notice about Professor Gause is that he is not dressed in the same fashion as his introducer, nor does he exude the same aura of organized efficiency. If anything, his aura is rather rumpled. He's wearing a wrinkled Harris tweed jacket, hiking shoes, and pants that look like he wore them in last night's tennis match and then slept in them. Nevertheless, he seems quite at ease and in control of the situation, with one exception.

He is scrutinizing the slide projector control as if it were a Venusian Death Ray Projector. Without looking up from his alien analysis, he announces, "It sure is nice to be here today." He then flicks up the first slide on a large screen directly behind the lectern.

You notice that the slide is ever so slightly out of focus, enough to be irritating, but not enough to provoke anyone into saying anything. Without looking at the slide, Gause says off-handedly, "I like to use this as my focus slide."

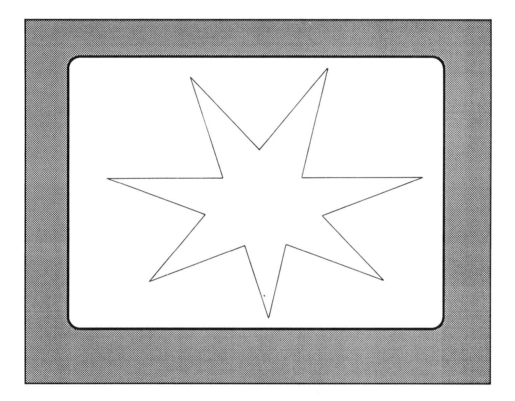

Figure 3-1. The focus slide.

You bend down to tie your shoe and when you look up, you notice he has moved on to the next slide. In excited tones, he begins, "This is a seminar about convergent design, which I define as 'a design process that consciously and visibly recognizes, defines, and removes ambiguity as effectively as possible.' "

At this point, Gause turns around and notices that the slide is out of focus. He has evidently mastered the slide control because he adroitly brings the slide into sharp focus. Whew! What a relief not to be eyeballing fuzzy slides all day! Even worse would be listening to a design lecturer who apparently didn't care about his own designs.

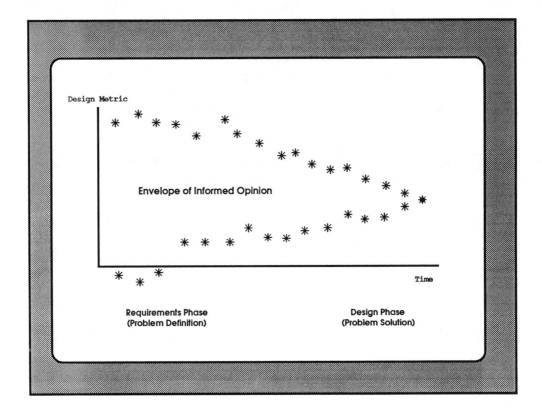

Figure 3-2. The Convergent Design Processes slide.

You now settle back, confident that everything is under control, as Gause uses a transportation device example to illustrate how to create an ambiguity metric. Then he has you break into groups to work on the example—a delightful respite from the lecturing you expected to go on all day. You're fascinated to notice that in your group the first estimates of market cost for a transportation solution range from $1 to $1 million. Then each group works through ten questions supplied by Gause, and the estimate range becomes $49.95 to $1 million.

You're having a pleasant time, and the lecturer's scheme for recognizing ambiguity seems reasonable enough, but things seem to bog down when he starts lecturing about some misty theory of design. Then you hear, "It's time for a required break, and we always meet requirements. After all, quality is meeting requirements, right?"

During the break, you share some tidbits with the other attendees and learn that everyone calls the speaker "Don." After you've milled around for about fifteen minutes, Don announces, "Your break is now one hundred percent complete.

Are you on schedule?" Everyone agrees, whereupon Don proclaims, "It's time to blast off in new directions."

3.2 A Test for Attentiveness

When everyone is seated, Don asks the class to answer the question posed on the next slide. "You are to work independently," he says, "privately writing your best estimate so as to make a firm commitment, and capturing your first impressions so you won't forget them when you hear other opinions." A slide appears:

How many points were in the star that was used as a focus slide for this representation?

> Reader: At this point, to get the full benefit of the exercise, write down your own answer without looking back at the star.

You're feeling suspicious and more than a little tricked, but you decide to comply when Don flips to a blank slide and says, "This question has everything in the world to do with design. For example, you might think of this as simulating a critical design decision that depends on a correct answer to the question. You might encounter this situation when an important event occurred that you did not realize was important at the time it happened. Everyone then has to recreate as much of the important information as possible. After you write down your answer," he promises, "I'll give you some specific examples. Now, please write down your answer to the question that was posed on the screen."

After you write down your answer, Don gives the promised example. "Remember Legionnaires' Disease in which many of the attendees of the American Legion convention at Philadelphia's Bellevue Stratford Hotel became ill some time after returning home? People died before a pattern was recognized and identified. Although an important medical event had occurred, nobody realized anything important was going on until some time later—just as you may not have recognized that the focus slide was important. When the importance finally was recognized, all that could be done was to reconstruct as much information as possible in seeking the cause and in designing the treatment and future prevention.

"There are many other examples," Don continues, "such as collapsing structures, penetration of security schemes, and accidents of all kinds, when the designers cannot duplicate the events they are trying to understand and to prevent with new designs. In fact, people were not even willing to go back to the Bellevue Stratford under any circumstances, and eventually the fine old landmark was torn down. That's why I've asked you to do the best you can, short of actually flipping back to the star slide."

A Test for the Reader

Reader: You have just simulated being at the first two hours of a typical one of our seminars. To enhance your experience, without looking backward or forward in the text, answer the following questions. Please answer them one at a time, in order, not looking at the next question until you have finished answering the previous question to the best of your recollection.

1. What do you think the answer is to the question posed on the screen?

2. With 100 people attending the seminar, how many different answers did the participants write down as their first-impression answer to the question?

3. What factors do you think are responsible for the differences among answers?

4. Write down, verbatim to the best of your recall ability, the question that you think you answered in question 1.

5. Write down the variants to the question that you think the seminar participants wrote when they were asked to recall the question that they thought they were answering.

Answer all the questions to the best of your ability before continuing.

3.3 The Clustering Heuristic

As you have probably guessed, this seminar actually took place. We provided this re-enactment to let you experience a few sources of ambiguity firsthand.

1. *What do you think the answer is to the question posed on the screen?* Only you know the answer to this, but keep it in mind as we discuss the actual seminar results.

2. *With 100 people attending the seminar, how many different answers did the participants write down as their first-impression answer to the question?* The 100 participants provided 18 different answers, as illustrated in Figure 3-3.

3. *What factors do you think are responsible for the differences among answers?* You can understand why other people's answers were the same as or similar to your original estimate, but you may be surprised by some of the answers in Figure 3-3. One way of under-

standing what went on in people's minds is to clump the results into clusters based on gaps in the histogram (see Figure 3-4). In our experience, there are four possible sources for these differences: observational, recall, and interpretation errors, plus problem statement ambiguity.

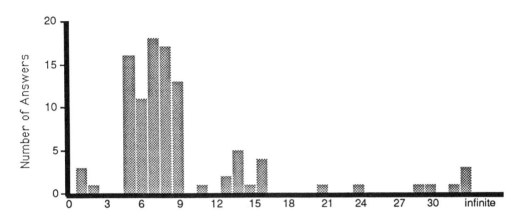

Figure 3-3. Histogram of answers to the "how many points" question.

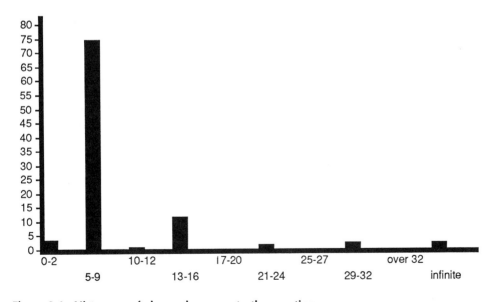

Figure 3-4. Histogram of clumped answers to the question.

3.3.1 Observational and recall errors

No two human beings can be expected to see things identically (observational error) or to *retain* what they did see identically (recall error). If you're not too egotistical, you can probably understand that the error may be yours as well as those who differ from you but share your cluster. Remember that the slide was deliberately shown in a haphazard manner, lasting for less than a minute at a time when people weren't paying much attention. Remember also that a lot of seemingly unrelated activity, including a fifteen-minute break, occupied the two hours between participants' seeing the slide and being asked the question. No one had any reason to believe that the slide would ever be considered again.

3.3.2 Interpretation errors

All this explains the spread within your own cluster, but how can we explain the extreme answers? If you were in the majority cluster, you thought the answer was somewhere between five and nine points. You can easily recognize a five-pointed star, but the star on the screen seemed more awkward and asymmetrical than that. Thus you reasoned it must have at least five points, and perhaps a few more. How could any intelligent, well-adjusted, well-meaning person earnestly believe that the star had only one or two points, or for that matter, 29, 30, or 32? Surely Don must have planted a few shills.

Could the reason be that the *question*, as asked, is subject to different interpretations? The three people who claimed that the star had an infinite number of points *must* have thought the star was composed of a series of continuous line segments. As high-school geometry teaches us, any continuous line segment contains an infinite number of points, so under that interpretation, these three answers were certainly correct and understandable.

Each cluster can similarly be explained by some other *interpretation* of the question. This then is a third major source of ambiguity, in addition to errors of observation and recall. Had there been only a single cluster containing all hundred responses, the interpretation ambiguity might have gone unnoticed. We notice differences in interpretation because of the different clusters, and each cluster may represent a different interpretation of the question.

Similarly, had the original image been clearly in focus and the subject of much attention at the time it was displayed, almost all of the variation would have been the result of differences in interpretation and recall, not observation. And, had the question been asked while the slide was still visible, recall errors would have been virtually eliminated, leaving only interpretation differences. The histogram would have shown no variation within clusters.

In other words, interpretation error tends to produce *separate* clusters, while observational and recall errors tend to manifest themselves as variations *within* each cluster. We can use this difference with the results of any ambiguity poll to help isolate the *sources*, rather than just the *quantity* of ambiguity.

3.3.3 Mixtures of sources of error

Of course, this way of interpreting clusters is not a hard-and-fast rule. Because the clustering itself is a visual, intuitive process, we may err in the way we define cluster boundaries. One participant in our sample said there were eleven points, and we decided to place that answer in a separate category because there were no answers of ten or twelve. But perhaps this person used the same interpretation as the 5-9 group, or the 13-16 group, but had poor recall or observation.

On the other hand, there could be two interpretations masked within the same cluster just because both happened to lead to the same estimate. Such different interpretations would go unrecognized by the clustering heuristic, though they may surprise us by resolving into separate clusters in a later poll.

The clustering approach must therefore be viewed as a useful indicator of ambiguity, rather than a scientifically guaranteed and validated means of defining its presence and isolating its source.

3.3.4 Effects of human interaction

In the seminar, we had the luxury of being able to ask the participants what they were thinking at the time they answered each question. In real development work, of course, this kind of questioning is not a luxury, but a necessity. The most important reason to tabulate the responses and visualize their spread in a histogram is not the histogram itself, but the discussion that follows. For example, after considerable discussion of possible interpretations of both histograms, we took another poll. This new poll revealed the results shown in Figure 3-5.

The first poll, insofar as people followed directions, was *independent*, though we must always be a bit wary of the idea of complete independence. Some people might have discussed the star slide during the break, and some might have actually peeked at their neighbors' answers. Given the structure of the exercise, such dependence effects are probably small.

Not so, of course, once the histograms had been posted and the various interpretations discussed. The histogram in Figure 3-5 shows a number of changes, the clearest being a shift away from the majority viewpoint (5-9) toward more extreme views (0-2 or infinite). In discussing the shifts, the participants confirmed that they had changed their answers because they had seen the problem in a different light. The change was due to new problem interpretations, not to altered recall or improved observation. Did you, in fact, consider changing your interpretation when you read this discussion?

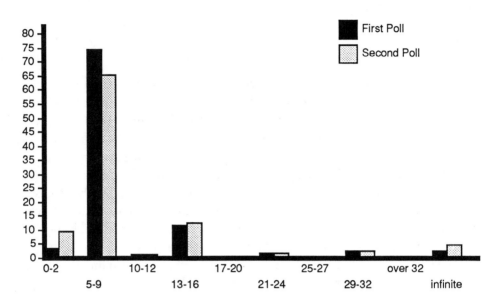

Figure 3-5. Clumped results of the second poll compared with the first.

3.4 Problem Statement Ambiguity

Now let's turn to the fourth question:

4.   *Write down, verbatim to the best of your recall ability, the question that you think you answered in question 1.* Again, only you know what you wrote down. Put this in front of you now as we discuss this and the fifth question.

5.   *Write down the variants to the question that you think the seminar participants wrote when they were asked to recall the question that they thought they were answering.* Recall that the exact question had been removed from the screen when the participants were asked, "This time, once and for all, please write down the answer to the question that was posed on the screen." Not having the question visible introduces yet another source of ambiguity: *problem statement ambiguity.*

We can easily imagine that without the question to refer to, not all participants were working on the same problem. Yet, until they were challenged by questions four and especially five, the participants failed to recognize the possibility of problem statement ambiguity. They were each quite content to launch into their own solution and, in some cases, to defend their solution to their fellow participants—totally unaware that they might be working on different problems.

At this point in the seminar, we collected the answers to question four. Here are the more common responses:

- How many points did the star have?
- How many points were there in the first slide?
- How many points did the star have in the first slide?
- How many points were on the star that was used as a focus slide?
- How many points were on the star in the first slide of this presentation?
- How many points were in the focus star at the beginning of this presentation?
- How many points were on the star that was used for focusing?
- How many points did the star used for focusing at the beginning of the presentation have?
- How many points (external) are on the star I used to focus?
- How many points were there on each star on the slide used for focusing?
- How many points did the star have that was used as a focus slide?
- How many points were on the star that was used to focus at the start of the class?
- What was the number of points on the star slide that was used to focus on?
- How many points were in the focus slide star?
- How many points in the picture of the star, used as a slide?
- How many points were there in the star on the slide?
- How many points were shown on the test star?
- How many points are on the star which was used as a focus slide?
- How many points were on the star slide used to focus at the beginning of the slide presentation?
- How many points did the star shown at the start of the presentation have?
- Determine how many points were present in the star shown earlier in the slide presentation?
- How many points were there in the original foil which was used to focus the foils?
- How many points were on the star that was shown at the beginning of the lecture?
- How many points were on the focus slide?
- How many points were on the star shown as the first slide?

You may be tempted to conclude that these questions are all more or less equivalent restatements of the same basic question. Certainly, in a seminar exercise, we can regard the whole question as trivial, but if we consider the exercise as a *simulation*—a model of the real world—then we cannot treat the exact wording so casually. In real situations, like the Legionnaires' Disease example, the questions are vastly more complex, and the situation far more serious. The exercise severely underestimates both the amount and importance of possible problem ambiguity in the real world.

Our problem statements must be precise, yet each variant statement of this relatively trivial problem *does* produce a different way of looking at the problem, which in turn produces a different solution. In real situations, differences as subtle as these spell the difference between a successful project and disaster.

3.5 Helpful Hints and Variations

• When working with requirements, use the recall heuristic directly. Simply take away the written requirements document and ask each participant to write down, from memory, what it said. Places where recall differs indicate ambiguous error-prone parts of the document. This is an important step because very few people will actually refer to the requirements document as they work, preferring to work with their memory of what the document says. Therefore, a document that is easy to remember correctly is much less likely to lead to design mistakes.

• Use the star exercise as a demonstration early in the requirements process to get people thinking in terms of ambiguity and their role in it.

3.6 Summary

Why?
The most important reason to tabulate the responses and visualize their spread in a histogram is not the histogram itself, but the discussion that identifies the sources of ambiguity present. Interpretation ambiguity tends to produce *separate* clusters, while observational and recall ambiguities tend to manifest themselves as variations *within* each cluster.

Attempting to recall the requirement or problem statement verbatim will reveal ambiguity, which must be removed before a successful requirement can be developed.

When?
Use the clustering heuristic with the results of any ambiguity poll to help isolate the *sources*, rather than just the *quantity* of ambiguity.

How?

1. Question participants about the interpretation of some part of the requirements document, and clump the results into clusters.

2. Analyze the clusters by asking people within each what they were thinking.

3. Isolate *observational error,* when people *saw* things differently, and *recall error,* when people did not *retain* what they saw. These may produce spread *within* a cluster.

4.   To account for spreads between clusters, ask participants to recall, without re-reading, the question they believe they were asked. This heuristic tends to spot *interpretation ambiguity.*

5.   After people discuss their observations, changes will be due to new problem interpretations, not to altered recall or improved observation.

Who?

Anyone who needs to understand the requirement, or any representative of a group of people who need to understand it, can usefully participate in this exercise to identify sources of ambiguity.

4 THE TRIED BUT UNTRUE USE OF DIRECT QUESTIONS

At this point, you may be skeptical and suspicious of the entire purpose of this book. "Sure," you may think, "removing ambiguity is important, even essential, to a successful product. But I already know how to remove ambiguity: Study the written requirements, then sit down with the client and ask some penetrating questions. In fact, if the client isn't easily accessible, I can send questions and get written answers. So, what's wrong with direct questions? Why do I need other techniques?"

The answer is quite simple: There's nothing wrong with direct questions. Indeed, if you ever hope to be a competent designer, you'd better master the direct question, the direct *observation*, and interviewing techniques generally. Those subjects, however, are well covered elsewhere,* so we'd rather use an illustrative problem to demonstrate why questioning without other tools and techniques is never enough to get the requirements right.

4.1 Decision Trees

One such tool for supplementing the direct question is the *decision tree model*. By picturing design as an exploration along the limbs of a "tree" of possibilities, this model helps you follow the results of getting your questions answered.

As shown in Figure 4-1, the root of the tree signifies the first vague statement of the problem, and that's the point where the exploration starts. Each branch of the tree represents a decision, and the leaves of the tree indicate all possible specific solutions. The exploration ends when you arrive at a single leaf. As you take one fork or another, you reduce the ambiguity in the problem and eventually arrive at a solution, a single leaf.

The further out you go along the decision tree's branches, the less ambiguity there is left to resolve. When you reach the final specific solution (leaf), all design decisions (branches) have been made, and there is no remaining ambiguity. In the

---

*For more on these subjects, see Parts III and IV, "Observation" and "Interviewing," respectively, in Gerald M. Weinberg's *Rethinking Systems Analysis & Design* (New York: Dorset House Publishing, 1988), pp. 43-86.

Vague Problem Statement

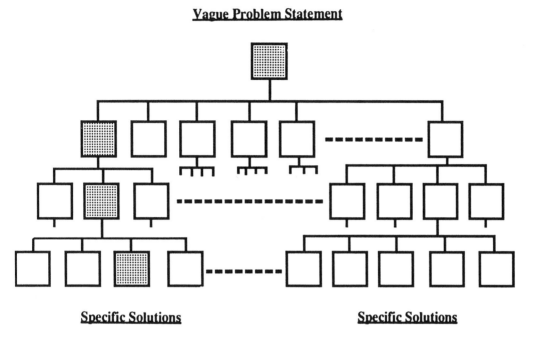

Specific Solutions   **Specific Solutions**

Figure 4-1. The decision tree is sometimes shown upside down, with the root (the vague problem statement) at the top and the leaves (specific solutions) at the bottom.

client's eyes, though, this solution may not be the *correct* solution. If you've made
even one false assumption (from the client's perspective), if you've made even one
incorrect branching decision, you will have solved the wrong problem.

Real design work, of course, doesn't follow quite this simple a model. We
often make mistakes in the design process (take wrong branches), but later discover those mistakes and correct them (go back and take the right branch). In this
model, mistaken assumptions don't necessarily lead to the *wrong* solution (wrong
leaf), but they do lead to *inefficient* design work (traveling along unnecessary
branches). By improving our ability to resolve ambiguity, we eliminate design mistakes, or at least improve the efficiency of our design work.

According to the decision tree model of design, any leaf on the tree other than
the precise one the customer wants is a design mistake. A more generous interpretation is that every leaf on the tree is a correct solution, but most of them are *correct
solutions to the wrong problems*. In other words, every leaf is a solution, but most
are *unacceptable* solutions.

4.1.1 Order of questions

We all know that real design work is not that black and white. In practice, some of the unacceptable solutions may be more acceptable than others. If the client wants a green bicycle, a purple bicycle would probably be more acceptable than a green skateboard. If that's the case, then "What color is it?" should probably be a late design question, and "Should it be a bicycle or a skateboard?" should be much earlier.

This example shows that the *order* of questions may in practice be just as important as the questions themselves. A sensible order will produce solutions closer to the "perfect" solution, and will also minimize the amount of work necessary to rectify incorrect assumptions.

This discussion suggests part of the answer to the skeptic's second question, "Why do I need other techniques?" You don't just need direct questions, you also need techniques to *help you ask your questions in the right order.*

Unfortunately, the false assumptions that lead to mistaken design decisions are more likely to be *perpetrated* near the root of the tree, but are more easily *recognized* near the leaves. That's why we need to exert more ingenuity and effort toward the resolution of ambiguity near the beginning of the design process, or, as we say in the jargon of development, we need more *front-end* effort, not just more *effort.*

Realistically, no amount of front-end effort will totally prevent erroneous assumptions. So, in design work, we'll need techniques to *simplify error correction.* When people have invested time, money, and ego in pursuing fruitless branches, correcting errors can lead to substantial political, financial, and psychological agony. Our techniques for exploring ambiguity will minimize such distress.

4.1.2 Traversing the decision tree: an example

To further explain the need for ambiguity-reducing techniques to supplement direct questions, let's consider a very ambiguous requirement. Imagine that we have founded a small design company, and our first client gives us our first requirements statement:

Design a transportation device.

This initial statement is at the root of the decision tree. Let's trace how our company arrives at a potential solution to this problem by following a series of direct questions through the decision tree toward one of the leaves. As each piece of information is supplied, we'll show how our position in the decision tree changes.

We start at the root, which defines the device's purpose as "transportation":

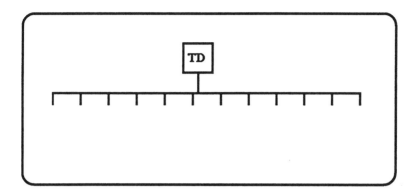

Figure 4-2. The root of the tree "Design a transportation device (TD)" branches out into various possibilities.

Designer: When is the solution needed?
Client: Within eighteen months (18).

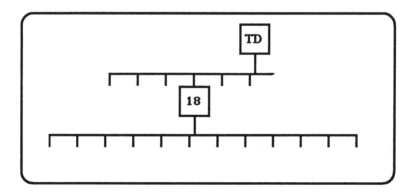

Figure 4-3. The answer to "When is the solution needed?" selects one of the limbs, which in turn has many possibilities.

Designer: What is to be transported?
Client: People (P).

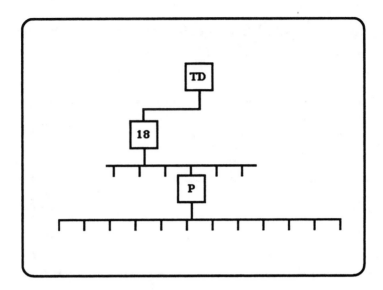

Figure 4-4. The answer to "What is to be transported?" selects another branch, again leading to many more possible branches.

Designer: How many people are to be transported at one time?
Client: One person (1).

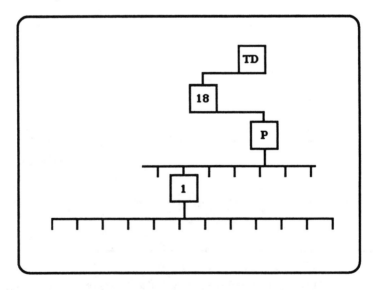

Figure 4-5. The answer to "How many people are to be transported at one time?" selects yet another branch, one level deeper in the decision tree.

Designer: How is the device to be powered?
Client: Only by the person being transported and natural forces that are readily available. It is to be "passenger-powered" (PP).

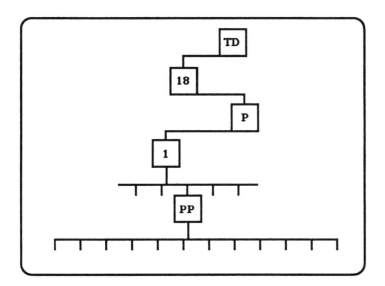

Figure 4-6. The answer to "How is the device to be powered?" causes the tree to grow deeper.

Designer: Over what kind of surface or terrain is the device to travel?
Client: It is to provide transportation over a very hard, flat surface (FS).

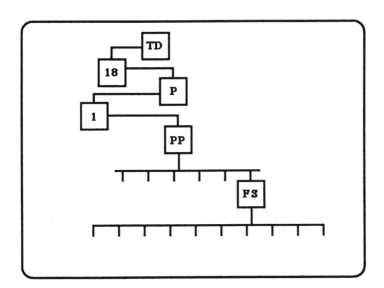

Figure 4-7. "Over what kind of surface or terrain is the device to travel?" Answer: "A hard, flat surface (FS)."

Designer: How far is the person to be transported?
Client: Up to 1.2 miles (1.2).

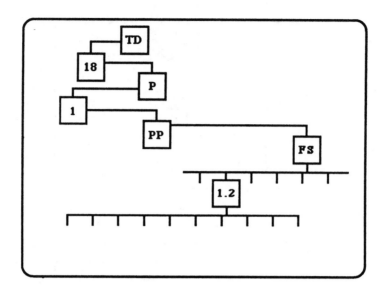

Figure 4-8. "How far is the person to be transported?" Answer: "1.2 miles."

Designer: How fast must the device travel?
Client: At least one mile every four hours (0.25 mph).

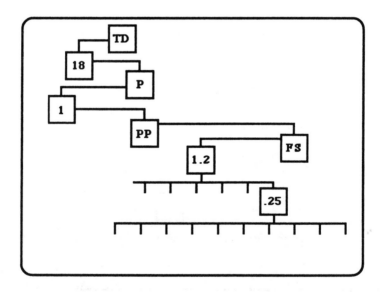

Figure 4-9. "How fast must the device travel?" Answer: "0.25 mph."

Designer: How reliable must the device be?
Client: It may not have more than one failure per thousand miles of operation, and that failure may not be one that endangers the passenger (1/1000, or .001).

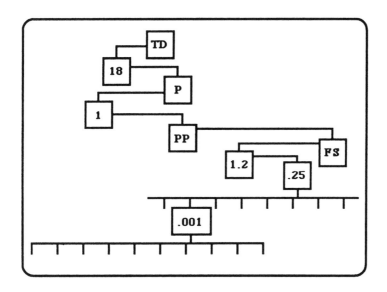

Figure 4-10. "How reliable must the device be?" Answer: "1 failure per 1,000 miles."

Designer: Are there any cost limitations?
Client: The device must be designed, developed, and manufactured so that it
            will cost less than three hundred Swiss francs each time the device is
            used (Sfr300).

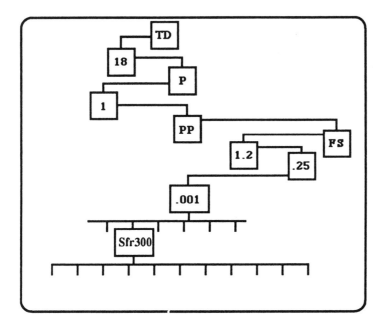

Figure 4-11. "Are there any cost limitations?" Answer: "300 Swiss francs per trip."

4.2 Results of an Ambiguity Poll

Was this a good series of questions? It should be, because this dialogue actually arose out of the questions raised by more than a thousand professional systems designers *after* they had initially estimated the cost of manufacturing "a transportation device." Their estimates reflected the ambiguity of this simple statement, covering the range of $10 to $15 billion.

After the questioning, the professional designers were again polled for the cost of manufacturing a transportation device. Their range of costs now narrowed from the original $10 to $15 billion range down to a range of $20 to $10,000. Clearly, their questions had lowered the requirements ambiguity by a substantial margin.

Were these professionals ready to start designing? When asked this question, they said, "Continue the dialogue."

We further asked, "How good is your grasp of the problem now?" Most of them felt quite confident that they knew what the problem was. When we examined the manufacturing cost estimates, we found two major clusters: $40 to $100 and $700 to $1,200. The designers in the lower cluster reported they were thinking in terms of devices like skateboards or roller skates. Those in the higher cluster had solutions in mind that resembled bicycles and tricycles. We labeled the two clusters *little wheels* and *big wheels.*

We can deduce from the small range of manufacturing costs that all of the designers assumed the device would be produced in numbers sufficient to minimize the impact of design and development cost. The number of devices to be produced had never been discussed, and the designers were not in communication with one another, so this ambiguity was evidently resolved implicitly by them all.

Although most designers were now thinking in terms of wheeled vehicles, the requirement for wheels had never been discussed either. Interviews with the designers showed that the requirement for providing transportation over a very hard, flat surface ruled out boats. For some people this ruled out airplanes as well, but a few designers saw that flying over the surface would also be transportation. When they thought of the self-propelled requirement, however, they dropped any idea of airplanes, although airplanes powered by the human passenger* have been widely publicized recently.

Once airplanes were eliminated, the very hard, flat surface suggested wheeled vehicles, though ice skates and sleds have been used for centuries and were certainly well known to the designers. These oversights suggest yet another answer to the skeptic's second question, "Why do I need other techniques?"

Just because we know something doesn't mean that we will make the connection between what we know and the requirements at hand.

---

*For an excellent history and description of human-powered airplanes, see M. Grosser, *Gossamer Odyssey* (Boston: Houghton Mifflin, 1981).

4.3 What Could Possibly Be Wrong?

Now, since all the designers were reasonably confident of their understanding of the requirements at this point, let's imagine that our little company did much the same, and proceeded with the design. Suppose we continued to move through the decision tree, going further and further "out on a limb." The final design was put on paper, the manufacturing process commenced, and the very first product has just rolled off the line. It is a compact, lightweight, folding tricycle that can be carried in a pedestrian's hip pocket.

At this point, our fledgling design company has a right to congratulate itself. It has solved its first design problem, and quite successfully. These compact folding tricycles satisfy the original problem statement and are completely consistent with all answers in the dialogue. We estimate that they can be profitably sold for not more than $49.95, with a sales volume of 100,000 per year for 5 years. Marketing estimates indicate that achieving this volume should be no problem. In fact, they should sell like hotcakes! What a slick solution!

But is it a slick solution to our clients' original problem? We arrange a meeting with our two clients, who are the mayors of the Swiss villages of Grindelwald and Wengen. When we demonstrate the fantastic folding tricycle, they express great delight with its portability. Even though we never asked them explicitly about portability, we congratulate ourselves for making a clever assumption.

Our reveries are interrupted, however, when Grindelwald's mayor asks, "I'm delighted with the portability of the device, but could you explain how it can be used to rescue a climber on the north face of Eiger Mountain?"

Evidently, we had missed something.

Our clients explain, what they had in mind was a *mountaineer's rescue device*, which is understandable since Grindelwald and Wengen base their economies on tourism in general and mountain climbing in particular. To keep tourism booming and to make sure climbers return, the villages' official guides must rescue stranded climbers. Attracting qualified guides willing to risk their lives for careless tourists is becoming harder and harder. Unless they can come up with a solution, the villages' economies are threatened.

Consequently, the mayors want a rescue device that they could require of all climbers attempting the mountain's five-thousand-foot vertical north face. Whenever a climber had to return due to bad weather, exhaustion, injury, fear, or inadequate supplies, the device would be put into use for the trip down the *hard, flat surface*. The devices could be rented in either village, and checked for safety after each rental. With a per use price of Sfr500, the villages would make a gross margin of about Sfr200 for each rental, in addition to saving lives, of course.

How embarrassing! Our ingenious solution was directed at the wrong problem: providing compact, highly portable *horizontal* transportation *across* hard, flat surfaces. The client requirement differed by ninety degrees: providing compact, highly portable *vertical* transportation *down* hard, flat surfaces.

4.4 Real Life Is More Real Than We Like to Think

Admittedly, this is a peculiar example. How many places in the world have hundreds of people trying to climb a mile-high vertical wall? But though this example was chosen to trick designers in a classroom situation, it was also chosen because it effectively simulates real-life ambiguities in a classroom-sized design project. In any real product development effort, the ambiguities may be less peculiar (though don't count on it), but they make up in quantity what they may lack in quality.

A relatively "normal" payroll package could have tens of thousands of assumptions lurking in its requirements. In one requirements review of a single eight-page piece of an on-line banking system, we turned up 121 ambiguities that were interpreted in at least two ways by different reviewers. This piece was only one out of 886 pieces. If we extrapolate the one review to the entire system, we get an estimate of 107,206 design decisions—which makes for an awfully big tree. To reach the right leaf without clarifying the requirements, each of those 107,206 decisions would have to be made by guess. What's the probability of guessing right on every one?

There is, of course, no realistic chance of guessing 107,206 decisions correctly, which relates to how we answer the skeptic's second question, "Why do I need other techniques?" Our "designers" assumed that the passenger was to be transported horizontally, or at most up relatively modest grades. This assumption was close to the root of the decision tree, which made later decisions erroneous or at least irrelevant. If our design company had been real, we would have lost a great deal of time and effort as a result, and probably a couple of customers besides.

The simple fact is that we need other tools and techniques for reducing ambiguity because *we, as normal human beings, are just not very good at seeing what we've overlooked.* For some reason, this psychological fact disturbs many designers, people who aren't disturbed at all by their inability to see X rays with their unaided eyes, for example, or hear ultrasound with naked ears, or hammer nails with their bare fists. Before embracing any new tools, we must learn to accept our shortcomings as perfectly human characteristics.

In sum, to guarantee that direct questions are successful, the questioner must be able to guarantee a perfect understanding of all earlier parts of the decision tree. This cannot be done with direct questions alone, so even if the direct question remains our *principal* tool, we need supplementary help to make our direct questions meaningful. And now that we've completed the introduction to this book, we're about to get some of that help.

4.5 Helpful Hints and Variations

* One of designers' major mistakes is to try to give customers what they *need*, rather than what they *want*. If you find yourself feeling that you know better what the customers need, don't assume you're smarter than they are. Instead,

try to convince your customers that they really need what you think they need. If you can't convince them, either produce what they *want*, or find yourself some other customer. It's not a good idea to work for people whose intelligence is so disparate from yours, in one direction or the other.

- If a product has unknown prospective customers, you won't be able to convince them of their need for the product at requirements time, because there is no product for them to see, and you don't even know who they are. Instead, you will have to add "Convince customers of the need for this product" to the requirements list. If the known current customers don't feel that this new requirement is needed, of course, you'll have to convince them first.

- When asking questions to determine requirements, use the decision tree directly. Each time a person answers a question, sketch the consequences, as we have done, onto a growing tree diagram, and show it to the person who answered. The question "Is this what you mean?" will often produce startling information about sources of ambiguity.

- We've convinced many skeptics of the limitations of direct questions through the transportation device exercise, and other similar stories. As you do requirements work, you'll have to convince other people who will typically accuse you of wasting time, when you do anything but ask direct questions. For your tool kit, collect a few vivid examples of climbing up the wrong branch, or even barking up the wrong tree, to counter such statements. The examples will be most effective, of course, if they come from the client's own organization but are not personally embarrassing.

4.6 Summary

Why?
You need other tools and techniques to supplement direct questions because a total direct question approach would require perfect human beings in order to be successful.

When?
Whenever you ask direct questions, you'll need tools to help you choose which direct questions are meaningful at that point in the decision tree, and to verify the meaning of the answers.

How?
The decision tree model itself is an outstanding tool to supplement direct questioning. Another tool is a supply of striking anecdotes to convince customers of the need to do more than ask them direct questions. (In subsequent chapters, we'll show other effective techniques.)

Who?
Anyone who is asking or being asked direct questions should understand human limitations, and know alternative ways to get information.

PART II   WAYS
TO GET STARTED

Barbara, Larry, and Todd from BLT Design have never seen a meeting room like the one they're in. It's more like a cave than a room, with curved chalk walls carved in bas-relief showing scenes that feature the great educators from Socrates to Sagan. There are four large blackboards, one on each side of the room—or, rather, one at each direction of the compass since the round room doesn't have sides. Byron, one of their hosts, motions them to be seated so that the meeting can begin and their other hosts, Wilma and John, also take seats.

Byron: We're here to begin development of a new product, which I call "Superchalk." John is here to help us with the user view. He teaches geometry and basketball at Consolidated Middle School. Wilma is here in a dual capacity. She's a professor of materials science at State College, so she's both an expert on the properties of chalk as well as a user of chalk. Are there any questions?

Barbara: John, should we be packing multiple colors of Superchalk in a single box?

John: Well, I don't know. I suppose that would be nice for geometry classes.

Larry: Are the requirements different for basketball than they are for geometry?

John: I never really thought of that. I don't think so, right off hand, but let me think about it.

Barbara: Good, we don't want to commit ourselves to any assumptions prematurely.

Todd: Wilma, do college students steal more chalk than high school students?

Wilma: (*haughtily*) *My* students don't steal chalk!

Todd: Well, I was reading that chalk has psychedelic properties, and that rubbing chalk into the skin is the latest rage among the college set.

John: Actually, I understood that it's more for body decoration than for any psychedelic effect. Our students model themselves after Australian aborigines.

Todd: Should we design Superchalk so that it can't be used on the body? For example, we could add an ingredient that stings when applied to the skin.

Barbara: The stinging ingredient could be mixed with the color. Do you think that would work, John?

John: I don't know. Wouldn't it sting my hands when I used it in class?

Larry: We could market a line of Supergloves to go with Superchalk.

Byron: Ladies and gentlemen, this is interesting chalk talk, but I wonder if we could start discussing my company's problem?

Barbara, Larry, and Todd, in unison: Oh, haven't we started already?

5   STARTING POINTS

If you're about to set out on some great exploration—like designing Superchalk—getting started can be the most difficult thing. All requirements work is preceded by some sort of an *initiation process:* Somebody has an idea that something should be designed and built. Regardless of its origin, the idea is the starting point for the requirements process.

Unfortunately, because people have their own individual ways of thinking, there can be many different interpretations or forms of the idea and hence many different starting points. This can prove disastrous for the requirements process.

If we're not careful, the initial form of the idea will start the whole process on a nonproductive track from which it may never recover. If participants don't begin thinking in concert, you'll lose them before you ever have them. As the evangelists say, you have to "preach to their condition." Once you have them in the tent, you may be able to convert them.

In this chapter, we'll look at the most common idea forms that give rise to different starting points, and discuss how to balance the need for a common starting point with the need for full participation from the outset.

5.1 A Universal Starting Point

How can we reduce the great variety of potential starting points to a single solid platform for exploring requirements? A possible solution is to regard every design project as an attempt to solve some problem, then reduce each starting point to a common form of problem statement.

A problem can be defined as

*a difference between things as perceived
and things as desired.**

---

*For a full discussion of problem definition, see Donald C. Gause and Gerald M. Weinberg, *Are Your Lights On? How to Know What the Problem Really Is,* 2nd ed. (New York: Dorset House Publishing, 1989).

Figure 5-1. A problem is best defined as a difference between things as perceived and things as desired.

This definition can serve as a template for measuring each idea for starting a development project. If the idea doesn't fit this definition, we can work with the originator to universalize the idea until it does.

5.2 Universalizing a Variety of Starting Points

Let's see how this universalization process can be used to reduce six different starting points to a common form of problem definition.

5.2.1 Solution idea

Perhaps the most common starting point is thinking of a *solution* without stating the problem that solution is going to solve—that is, what is *perceived* (and *by whom*) and what is *desired*. Here are a few examples we've experienced.

A marketing manager told a systems analyst, "We need sharper carbon copies of our sales productivity report." Rather than immediately begin a search for
a way to produce sharper carbons, the analyst asked, "What problem will sharper
carbons solve for you?" The manager explained that the carbons didn't make very
good photocopies, so the salespeople had trouble reading them. "So," the analyst
confirmed, "you need one clear copy for each salesperson, and you're now making multiple copies of the report we give you?" Eventually, through such give and
take, the problem was redefined as a need to provide timely and clear comparative
information to a sales force of four hundred—something that was easily accomplished by a slight modification to an existing on-line query system. The final design didn't have sharper carbons; it didn't have carbons at all, or even paper.

In another case, a university dean said, "We need a way to attract more students." The dean never said *why* they needed more students, and each faculty member hearing the statement formed a different idea. Some thought "more students"
meant getting more outstanding students. Some thought "more students" meant
being able to support more teaching assistants in certain departments. Still others
thought "more students" meant the dean wanting to fill the vacant dormitory space.

After arguing for months about the best way to get more students, the faculty
finally learned what the dean really wanted: to create the impression in the state
legislature that the school was doing a higher quality job by *increasing the rejection
rate of applicants,* so the university appropriation would increase. Once this goal
was understood, the faculty approached a solution in several ways, none of which
involved an increase in student enrollment.

5.2.2 Technology idea

Sometimes we don't have a problem in mind, at all, but literally have a solution
in hand: a solution looking for a problem. Have you ever felt when tearing off
those perforated strips on computer paper that there ought to be something useful
to do with them? The perforated strips are the solution, and the problem is "What
can we use them for?" After thirty years of searching, Jerry bought Honey, a German shepherd puppy, and suddenly he discovered the problem his solution was
looking for. Computer paper edges, crumpled up, make perfect litter for puppy
nests!

When a new technology comes along, it is often a solution looking for a problem. The Post-It™ note developed by 3M is a conspicuous example. The semi-
stickiness was originally just a failed attempt to produce an entirely different kind
of adhesive. Instead of simply discarding it as another failed project, the 3M people thought of problems for which such semi-adhesive properties would provide
a solution. They created Post-It notes, but the solution-to-problem process didn't
stop there. As soon as Post-It notes appeared in offices, thousands of people began seeing problems that they would solve.

Some of the high-technology companies we work with are dominated by this kind of solution-to-problem starting point. In effect, their problem takes the form of the following perception and desire:

- *Perception:* We own a unique bit of technology, but others don't want to give us money for it. For example, Byron's chalk company buys rights to a new vein of chalk that has exceptional purity and strength. To most people, however, chalk is just chalk.

- *Desire:* Others will pay us a great deal of money for the use of this technology in some form. For instance, if Byron can create the idea of Superchalk in the public mind, the unique purity and strength become an asset of increased value.

Such a problem statement allows the technology to become a kernel around which many designs can be built. Without it, technology firms often make the mistake of believing that "technology sells itself." Although this may be true in certain cases, usually it's an after the fact conclusion. Ordinarily, it requires an enormous amount of requirements and design work to turn a solution into a problem that will require it—for example, to make teachers believe that they can't really teach well without Superchalk.

5.2.3 Simile

Many product development cycles start with a variety of metaphorical thinking—a simile, or comparison, as when someone says, "Build something like *this*." Although the customer may emphasize "this," the job of the requirements process is to define "like."

For instance, Maureen, the leader of a software project, told her team she wanted a new user interface that was "like a puppy." "First of all," she elaborated, "people see a puppy and are immediately attracted to it. They want to pet it, to play with it. And they aren't afraid of the puppy, because even though it might nip them, or even pee on them, puppy bites aren't serious injuries, and puppy pee never killed anyone. Also, you can't really hurt a puppy by playing with it."

Although the team couldn't yet build a system with this requirement, playing with the simile did inspire them to ask probing questions. "How about housebreaking and obedience training for the puppy?" a teammate asked. Maureen thought a bit, and said, "Yes, the interface should be trainable, to obey your commands, so it becomes your own personal dog."

"Okay," asked someone else, "will it grow up to be a dog, or remain a puppy?"

"That's easy," she said. "It will stay a puppy if you want it to be a puppy, but if you prefer, it will grow up to be a real working dog that does exactly what you say."

"What kind of working *dog?*"

"A watchdog, for one thing. It should warn you of dangerous things that might happen when you're not paying attention."

Someone else got into the spirit by asking, "What about a sheepdog? It could round up the 'sheep' for you, and put them safely in the pen. And guard them from anyone stealing any."

By this time everyone was involved, and the requirements process was running like a greyhound, though not necessarily in a straight line, as when someone asked, "How about fur? Should it be a longhair or a shorthair?" Nobody could figure out what that meant, though it did lead to an extensive discussion of touch screens and other interface hardware that they had never previously used. Eventually this tangent was clipped by someone observing, "Our tail is starting to wag the dog."

The simile is very good as an idea-generation tool, but eventually the requirements group has to groom their ideas into prize-winning form, and that requires some idea-reduction tools. You know when the simile has become a bit dog-eared when you can no longer make fruitful connections between it and your product. It's important, though, to keep it going a bit past the point where it becomes ridiculous, just to be sure you've generated enough ideas.

5.2.4 Norm

Many people do not consider themselves metaphorical thinkers, believing they think more concretely. In the requirements process, they would more likely say, "Here is a chair. Design a better chair." or "Here is ordinary chalk. Design a superchalk." In fact, *the norm is also a metaphor,* but one that seems literally "close" to the thing desired. The great danger of using a norm is the constriction on our thinking once we identify what would almost satisfy the customer.

Another great danger is making one big leap in logic to the end result. Instead, starting with a norm and working by increments tends to protect us from the colossal blunder. The Wright brothers, for instance, were bicycle builders, and they used many of the norms from bicycle construction to create their success at Kitty Hawk.

A third danger is starting with the wrong norm, which could prevent us from making a great leap forward when one is possible. Orville and Wilber Wright did use a rail to launch their plane, but they didn't become the first heavier-than-air fliers by putting wings on a locomotive (Figure 5-2).

5.2.5 Mockup

If we decide to use a chair for a norm, though, your mental picture of a chair may be very different from mine. A mockup is a way to protect against this ambiguity, by providing an actual scale model of a product. Moreover, we can benefit by using the mockup to demonstrate, study, or test the product long before the product is actually built.

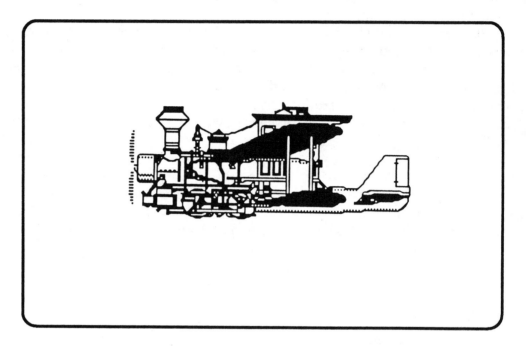

Figure 5-2. Don't let the norm dictate the form. If the Wright brothers had been train builders, they might have specified a plane that looked like this hybrid, which might have been on the right track, but would have had a hard time getting off the ground.

A mockup serves as a norm, when no norm exists, or when none is available. As such, it has all the advantages and disadvantages of a norm. It also has the advantage and disadvantage of being a fantasy product. When we use a mockup, we aren't restricted to what exists, but on the other hand, we can easily mock up a product that could never actually be built.

In printing, and in computing, for example, the mockup is often in the form of a layout of printed matter, or material on a screen. The customer and users can point to the layout and say, "Yes, that's what I want," or "No, what's this doing here?" What we are actually testing with the mockup is their emotional response—their desire. In effect, a mockup says, "This is what we think the product's face will look like. Let's see how you react to this!"

5.2.6 Name

Many ideas for design projects simply begin with a name: Create Superchalk. Build me a table, chair, pencil, clock, elevator, steering wheel, speedometer, or bicycle. Although the name provides a quick and common connection for all participants to grasp, names also come with a large baggage of connotations. As we've seen, each word is worth a thousand pictures, and each connotation of a name may introduce implicit assumptions.

For instance, one of the reasons Jerry spent thirty years searching unsuccessfully for a use for computer paper edges was that the name itself narrowed his thinking unnecessarily. Our colleague Jim Wessel observed that his four-year-old daughter isn't so limited. She cuts these strips into smaller pieces and calls them "tickets." We would do well to emulate the four-year-olds who have little trouble making up names for objects that don't have them already.

5.3 The Can-Exist Assumption

So where have we come in our search for a universal way to start a project? After exploring solution ideas, technology ideas, similes, norms, mockups, and names, we seem further away from the universal starting point we are seeking. But if we look back over these diverse starting points, we find that there is one thing they all share—the *assumption of existence.*

For a development project, existence is the number one assumption because what we want to develop, by definition, doesn't yet exist. Yet what do all these starting points have in common? They are ways of *pretending that a successful product is possible.* People would never begin a project unless they assumed that at least one solution *can* exist.

Restated, here's our universal starting point (Figure 5-3):

All development starts with the assumption that a solution to a problem can exist.

The remainder of requirements work can be seen as a clarification of that assumption. As we proceed through the book, we'll show you a series of tools and techniques that start with "might be," the *assumption* of existence, and work through one assumption at a time to "is," the *reality* of existence.

5.4 An Elevator Example

Let's set the stage for a single product development example that we'll use to provide continuity in the following chapters. For this example, we'll walk through the early stages of the design activity to illustrate what we mean by exploring requirements:

- Imagine that you have been selected to participate in a two-day session to seek dramatic new directions in the technology of control, information development, and display. The output of those two days is to be a solid requirements document, with as much ambiguity as possible exposed and removed.

Figure 5-3. All development starts with the assumption that a solution to a problem can exist.

- During these two days, you are to develop concepts for a control and information display device for elevators used in high-rise buildings from ten to two hundred stories tall. The device must provide information and control to any of these floors as deemed appropriate by the building's management.

- The device will incorporate state-of-the-art sensing, control, and information transfer and display technologies and must be entirely realizable, in prototype form, within six months. It must be available from a stable, fully developed manufacturing facility within eighteen months. The device is to be uniquely functional such that it will capture the elevator control and information display market for the foreseeable future.

5.4.1 Naming our project

If we're going to talk about our project, we'll need a clear, unambiguous name. In Chapter 12, we'll develop a *naming heuristic* to do just that. For now, let's just apply this heuristic, because if we don't give it a name, somebody else will. Later, we'll see why the heuristic works, and why it is valuable.

Suppose the first name we think of is "Elevator Control Device." Here is a list of reasons for why this is not a good choice:

- We may want to do more than *control* the elevator.

- We may want to control something *other* than the elevator. We may want to control *people* in and around the elevator, or control the *environment* of the elevator.

- Nothing in this name implies information display, yet this seems to be an important function according to the problem statement.

After considering these shortcomings, we offer a second name, "Elevator Display Device." This name has the following difficulties:

- The name strongly implies *visual* transfer of information, and we don't want to limit our thinking this early. We may want to impart information through a variety of sensory channels, such as sound or smell.

- The word "display" may also be interpreted to imply a traditional "screen" approach.

- This device, according to the problem statement, is to facilitate the flow of information from inside the elevator out (display) as well as from the outside of the elevator in (control).

Having tried these two names and found them wanting, let's go with "Elevator Information Device" as the working title. We could try for an even better name, but we're eager to embark on our explorations of other requirements tools. As you read these chapters, notice how your understanding of the Elevator Information Device Project is transformed by each exploration method. When we get to the naming heuristic in Chapter 12, try to apply this understanding to produce a better name.

5.5 Helpful Hints and Variations

- By the time you realize a project has begun, it's usually well past the true beginning. The moment you realize that you're actually starting to develop something, stop and write down everything you can recall that led up to this beginning. Then interview others and find out what they recall. It's in those pre-beginning thoughts that the biggest assumptions are made.

- A good tool, once you've begun in a particular way, is to play at "restarting" the project from each of the other beginnings. For instance, if you started with a name, make a mockup, a metaphor, or a norm, and act as if you were starting fresh.

- It's not literally true, of course, that all development starts with the assumption that a solution to a problem can exist. For instance, research projects are labeled " development" in order to get financing. Perhaps that's the true difference between research and development—the assumption that a product will certainly result. Of course, it's only an *assumption*, as proved by the data showing that fewer than one-third of development projects actually produce a viable product. So take a little time as you're starting to figure out whether you're working on a research project or a development project.

- Some projects begin with a "feasibility study," to try to test the assumption of existence. A feasibity study must then involve the root and the first few branches of the design tree. So, in fact, a feasibility study is another name for "early requirements definition."

5.6 Summary

Why?
Pay special attention to how a project begins because that's where the design tree is rooted. Right at the beginning is the easiest place to slip into major false assumptions, so learn to develop the habit of scrutinizing your starting point.

When?
It's never too early to begin looking at the beginning. In fact, by the time you realize that you've begun, the true beginning has already ended. How do you know when the beginning has ended? As soon as you talk as if there is no question of the existence of a solution.

How?
The fundamental idea is to slow down the beginning, to see requirements, and think about requirements, as the Zen priests say, with a "beginner's mind" for as long as possible. To the beginner's mind, all things that exist are equally, astonishingly, improbable.

Who?
Locate everyone who was involved at the beginning of the project.

6 CONTEXT-FREE QUESTIONS

Now that there is a starting point and a working title, our next step in the project is invariably to ask some *context-free questions*. These are high-level questions that can be posed early in a project to obtain information about *global* properties of the design problem and potential solutions. Context-free questions are completely appropriate for any product to be designed, whether it's an automobile, airplane, ship, house, a jingle for a one-minute television commercial for chewing gum, a lifetime light bulb, or a trek in Nepal.

In terms of the decision tree, context-free questions help you decide which limb to climb, rather than which branch or twig. Because context-free questions are independent of the specific design task, they form a part of the designer's toolkit that can be used before getting involved in too many enticing details.

6.1 Context-Free Process Questions

Some context-free questions relate to the process of design, rather than the design product. Here are some typical context-free process questions, along with possible answers for the Elevator Information Device Project. To appreciate how these questions can always be asked, regardless of the product being developed, also try answering them for some current project of yours, like "Trekking in Nepal."

Q: *Who is the client for the Elevator Information Device?*
A: The client is the Special Products Division of HAL, the world's largest imaginary corporation.

Q: *What is a highly successful solution really worth to this client?*
A: A highly successful solution to the problem as stated would be worth $10 to $100 million in increased annual net profit for a period of five to ten years. The product should start earning revenue at this rate two years from now.

Q: (Ask this if the answer to the previous context-free question does not seem to justify the effort involved.) *What is the real reason for wanting to solve this problem?*

A: The Elevator Information Device Project is a pilot effort for a range of commercial (and possibly even home or personal) information transfer devices. If we can demonstrate success during development and early marketing phases, this project is expected to spawn an independent business unit with gross revenues in seven years of $2 billion per year.

Q: *Should we use a single design team, or more than one? Who should be on the team(s)?*
A: You may choose whatever team structure you desire, but include someone on the team who knows conventional elevator technology, and someone who understands building management.

Q: *How much time do we have for this project? What is the trade-off between time and value?*
A: We don't need the device before two years from now because we won't be ready to market it, but every year we delay after that will probably reduce our market share by half.

Q: *Where else can the solution to this design problem be obtained? Can we copy something that already exists?*
A: To our knowledge, nowhere. Although many solutions exist in the form of control and information display panels for elevators, the approach adopted here should exploit the latest in sensing, control, and information display and processing, so copying doesn't seem appropriate. We have no objection to your copying features that exist elsewhere, and you should be aware of what else has been done in the field, so you can surpass it.

6.2 Potential Impact of a Context-Free Question

Are context-free questions really worth such a fuss, and why do they need to be asked so early? Let's look at an extreme but real example of an answer to the second question, *"What is a highly successful solution really worth to this client?"*

At 3 A.M., a man in dirty jeans and cowboy boots showed up at the service bureau operation of a large computer manufacturer. Through the locked door, he asked if he could buy three hours' worth of computer time on their largest machine that night. The night-shift employees were about to turn him away when one of them said, "Well, it costs $800 an hour. Is it worth $2,400?"

"Absolutely," said the cowboy, who emphasized the urgency by pulling a large wad of $100 bills from his pocket and waving them at the employees on the other side of the glass door. They let him in, took his payment in cash, and let him run his job on the machine. It turned out that he owned a number of oil wells, and as a result of his computations, and especially the courteous treatment he received, he *bought* three of the giant machines, at a cost of some $10 million. If the employees had *assumed* the answer to the "What's it worth?" question based on his appearance, there would have been no sale (Figure 6-1).

Figure 6-1. **A little manure on the boots may disturb city folks, but in requirements work, you learn not to mistake appearance for value.**

6.3 Context-Free Product Questions

Some context-free questions relate to the design *product,* rather than the design process. Here are some typical context-free product questions, along with possible answers for our Elevator Information Device Project. Notice again how these questions can always be asked, regardless of the process or the particular product being developed.

Q: *What problems does this system solve?*
A: This system is to make riding an elevator enjoyable, safe, and informative. It is to enable riders to reach any desired destination in the building, and to assist riders in being prepared when they reach their destinations. It is to minimize all unlawful acts for which elevators are frequently used, including mugging, assault, robbery, and trespass, such as door-to-door magazine sales.

Q: *What problems could this system create?*

A: If we are not careful, the system could create an unusually heavy demand for elevator space by extra riders traveling for sheer pleasure, safety, and information. Also, when we provide critical information, we expose ourselves to possible litigation when the information is incorrect or misleading. Finally, the cost of producing elevators may grow as riders become ever more demanding of the quantity and quality of information.

Q: *What environment is this system likely to encounter?*
A: This system is to be installed in single-purpose buildings such as offices, apartments, and hotels, as well as multipurpose buildings, such as "total living" buildings that contain apartments, stores, offices, hospitals, and other services required for self-contained living. The system is to be installed in the usual passenger and freight elevators, and be operated by the same population that now operates such elevators.

The Elevator Information Device is also expected to be used on the elevators during construction of new buildings. At this time, only construction workers with appropriate credentials will be allowed to use the elevators. These credentials will, of course, vary from floor to floor as the construction progresses.

As the elevators may be subject to improper use, loading conditions will be monitored at all times and unsafe loads will trigger special actions and warnings.

Q: *What kind of precision is required or desired in the product?*
A: To the extent that this is an information device, and not a control device, the precision requirement is not severe. If the system gives misinformation no more than once in a hundred times, that would be acceptable, as long as people could recover from the error. On the other hand, some functions of the system are directly involved in public safety, and these must meet much more demanding precision requirements, perhaps fewer than one error in ten thousand interactions.

6.4 Metaquestions

A third broad category of context-free questions is metaquestions, or questions about questions. Here are some typical metaquestions, along with possible answers from the Elevator Information Device Project.

Q: *Am I asking you too many questions?*
A: It's fine so far, but I do have other work, so I'll let you know if I have to stop.

Q: *Do my questions seem relevant?*
A: Most of them.

(If the answer had been negative, follow up with the metaquestions: *Which ones? Why?* These questions invariably reveal major misconceptions.)

Q: *Are you the right person to answer these questions? Are your answers official?*
A: Yes, I'm the one who knows the most about this, but I do have to get my boss to sign off on anything before it's official. *(Be careful here. Many people will be reluctant to answer this one accurately. Do you always like to admit that you're not an authority or that you're not in charge?)*

Q: *In order to be sure that we understand each other, I've found that it helps me to have things in writing so I can study them at leisure. May I write down your answers and give you a written copy to study and approve?*
A: Actually, I'd prefer to write them and send them to you for approval. Is that all right? *(Again, this can be a sensitive question for some people. It's not important who writes things down, but somebody should do it, and the other party should check it to ensure clear communication.)*

Q: (Use this if your communication has been only in writing so far.) *The written material has been helpful, but I find that I understand some things better if I can discuss them face to face. Can we get together at some point, so we can know each other better and can clarify some of these points?*
A: Sure. How about lunch tomorrow, then we'll spend the afternoon in my office.

Q: *Is there anyone else who can give me useful answers?*
A: Tomorrow after lunch, I'll introduce you to a couple of specialists from my organization who may be able to give more details than I can on some points.

Q: *Is there someplace I can go to see the environment in which this product will be used?*
A: Our office building is actually a good example. Why don't we start there tomorrow, when you're here?

Q: *Is there anything else I should be asking you?*
A: Well, there are a lot of details still to be covered, but I assume you'll get to them later. *(This is a question you can ask at the end of every interaction.)*

Q: *Is there anything you want to ask me?*
A: I'd like to hear a little more about your company, and about your personal background, so I'll know better what I can assume you know and don't know. There may be other things, too. I'll think about it and get back to you. *(This is another question you can ask at the end of every interaction.)*

Q: *May I return or call you with more questions later, in case I don't cover everything this time?* (Figure 6-2)
A: Sure. Here's my phone number. Just don't expect to find me on Monday mornings or Friday afternoons. *(This is a question you should definitely ask at the end of every interaction.)*

Figure 6-2. Perhaps the most important question is, "May I return or call you with more questions later, in case I don't cover everything this time?"

6.5 Advantages of Context-Free Questions

From examples of context-free questions, we can see how the answers have greatly increased our understanding of the Elevator Information Device Project, with only a small investment of effort. A great inherent advantage of context-free questions is that they can be prepared in advance, before much is known about the project. In addition, they help get over the awkwardness of starting a new project and new relationships.

Later on in the process, context-free questions focus attention on global issues concerning the product, as well as on issues in the design process itself. The answers to the metaquestions, in particular, help keep everyone on track in developing requirements. Metaquestions raise the kinds of issues that are most likely to be resolved implicitly, and thus ambiguously. In product areas with ancient histories, such as printing, home building, and highway construction, many metaquestions are answered by tradition. Unless we address these issues explicitly, we're likely to hear at the end of a project, "Oh, we thought you *knew* that. We always do it that way." (Figure 6-3)

Figure 6-3. Context-free questions will help you avoid hearing this: "Oh, we thought you knew that. We always do it that way."

6.6 Helpful Hints and Variations

● Here is a list of context-free process questions that we have found widely useful:
Who is the client?
What is a highly successful solution really worth to this client?
What is the real reason for wanting to solve this problem?
Should we use a single design team, or more than one?
Who should be on the team(s)?
How much time do we have for this project?
What is your trade-off between time and value?
Where else can the solution to this design problem be obtained?
Can we copy something that already exists?

● Here is a list of useful context-free product questions:
What problems does this product solve?
What problems could this product create?
What environment is this product likely to encounter?
What kind of precision is required or desired in the product?

● The following metaquestions have been used with great success:
Am I asking you too many questions?
Do my questions seem relevant?
Are you the right person to answer these questions?
Are your answers official?

In order to be sure that we understand each other, I've found that it helps me to have things in writing where I can study them at leisure. May I write down your answers and give you a written copy to study and approve?

- Use this if your communication has been only in writing so far:
The written material has been helpful, but I find that I understand some things better if I can discuss them face to face. Can we get together at some point, so we can know each other better and can clarify some of these points?
Is there anyone else who can give me useful answers?
Is there someplace I can go to see the environment in which this product will be used?

- The following metaquestions should be asked at the end of every interaction:
Is there anything else I should be asking you?
Is there anything you would like to ask me?
May I return or call you with more questions later, in case I don't cover everything this time?

- Context-free questions can tease out essential information about the most subtle assumptions. You may easily miss important information that comes in the form of nonverbal reactions such as hesitations, puzzled or delighted looks, amusement, anger, and changes in posture. Whenever possible, use two people to conduct the context-free questioning, one to ask and one to observe and note nonverbal reactions. The observer can ask metaquestions of the type:
I notice that you hesitated a long time before answering that question. Is there something else we should know?

- Context-free questions often reveal conflicting assumptions among various interested parties. Because you have prepared the questions in advance, you can more easily compare replies, and you should always do so. Then you can ask:
When I asked X about that, she said Y. Do you have any idea why she might have said Y?

- An easy way to compare replies is to interview two or more people at the same time. If they have differences in their assumptions, you may see a clear reaction. Then you can ask the following metaquestion:
I notice that you don't seem to agree with that reply. Would you tell us about that?

- When interviewing two or more people together, however, be on the alert for one person's answers inhibiting another's. Such inhibitions are going to haunt you as you proceed through the project, so you may want to take this early opportunity to ask:
Are you comfortable with the process right now?
Is there any reason you don't feel you can answer freely?

Of course, if they don't feel comfortable, they may also not feel comfortable *saying* they're not comfortable. If you sense that this situation may exist, try asking the previous questions in private, of each participant.

- All of this leads to another group of tough metaquestions that are critical, but that cannot always be asked directly:

  What can you tell me about the other people on this project?
  How do you feel about the other people working with us on this project?
  Is there anybody we need on this project whom we don't have?
  Is there anybody we have on this project whom we don't need?

  If a direct approach is not suitable, handle such questions not by asking them, but by looking for their answers buried in the answers to other questions. When someone mentions another person, make a note of the context and, possibly, ask,

  Can you tell me more about that person?

6.7 Summary

Why?
Context-free questions force you to look at the global issues in the design process, to put you on the right limb, rather than out on a limb. Since they are appropriate for any design project, they can be prepared in advance and used from one project to another.

When?
Context-free questions are used very early in the requirements process, before specific decisions get locked in.

How?
Follow this procedure:

1. Once an atmosphere of trust and rapport has been established, explain that you want to ask some very general and searching questions. Explain the importance of this type of question, and be sure you get agreement on the value of the process at this time.

2. Some people have trouble understanding why such general questions are useful. If necessary, go very slowly. Ask one question, explore the answers thoroughly, then explain how the information you obtained was useful to you.

Who?
Context-free questions are particularly useful with people who have broad knowledge of the present product or future requirements, knowledge that they may not realize they possess, or think is important.

7 GETTING THE RIGHT PEOPLE INVOLVED

Perhaps the most common single mistake in development efforts is to leave an essential person out of the process. That's why several context-free questions are designed to tease out information about people whose input will be useful:

> *Who is the customer?*
> *Who should be on the team?*
> *What problems does this product solve (that is, for whom)?*
> *Are you the right person to ask these questions? Who is?*

7.1 Identifying the Right People

The last of these questions is really the key, but you'll seldom find a person who knows the complete answer. Therefore, tools are needed to identify, locate, and gain the participation of all the right people.

7.1.1 Customers versus users

The very first context-free question is always "Who is the customer?" (In some businesses, we would ask about the "client"—we'll use the two terms interchangeably.) Why is this the first question? Because the customers are the ones who *pay us for the requirements work*. And, if nobody is paying, there is no work to do.

Product development is not possible without a customer, but you may be your own customer. That is, you may be paying for the work yourself, either as a volunteer or as an entrepreneur. In either case, the customer ultimately pays, and the customer ultimately makes the final decisions.

A *user* is any individual who is *affected* by, or *affects*, the product being designed. In the Superchalk Project, the people from BLT Design weren't paying very careful attention to the distinction between user and customer, and it caused trouble in their first meeting. Byron, who is the company's president, is the customer. Byron invited Wilma and John to the meeting because as teachers, they represent a major part of the user community. Wilma is also another kind of user—a "subject matter expert"—because of her expert knowledge of the properties of chalk-like materials.

Customers are obviously users, too, because paying for the product certainly must affect their bank accounts. But not all users are customers. For example, consider the toy business. In 1945, Jerry's dad, Harry, went into business with a line of wooden toys that had been user-tested on hundreds of children at Northwestern University. Unfortunately, the toys failed to sell well because in 1945 children were toy *users*, but not toy *customers*. The parents not only did the paying, they did the choosing. Assuming there was no difference between users and customers turned a design success into a business failure.

Two generations later, the situation has changed radically. Parents still pay, so in that sense they are still customers, but they no longer do most of the choosing. Television advertising and parental permissiveness have placed most of the toy decisions in the hands of the children. Children have become not only users, but customers as well. Many traditional toy manufacturers have gone out of business because they failed to notice this change in the customer/user relationship.

7.1.2 Why include the users?

It's obvious why you have to identify the client, first thing. The client makes the external decisions, like how much the product is worth. If nothing else, the client hires and fires the designers. But why include the users in the requirements work? Wouldn't life be simpler if you could just give them what you know will be good for them?

Yes, life would be simpler—until it's time to put the product into use. The users are the people who will make or break the product. For instance, our clients' studies of software tools indicate that seventy percent of tools that are purchased are never used. Moreover, of those tools that are *used*, ninety percent are used by just one person or small group, even when purchased for a large organization.

In our experience, this waste of money starts when requirements for tools are drafted by a group that excludes the targeted users. Not only don't the tools meet requirements, but the potential users feel no involvement in the process, and thus no commitment to give the tools a good try. And, if this example doesn't impress you, just think of all the Christmas presents that are never removed from their boxes before being exchanged.

7.1.3 The Railroad Paradox

When trying to find all users, we need to beware of the Railroad Paradox.* When railroads are asked to establish new stops on the schedule, they "study the requirements," by sending someone to the station at the designated time to see if some-

---

*For a fuller development of the Railroad Paradox, see Gerald M. Weinberg, *Rethinking Systems Analysis & Design* (New York: Dorset House Publishing, 1988), pp. 56-59.

one is waiting for a train. Of course, nobody is there because no stop is scheduled, so the railroad turns down the request because there is no demand.

Figure 7-1. The Railroad Paradox: Don't expect users to be waiting for trains that aren't scheduled yet.

The Railroad Paradox appears everywhere there are products, and goes like this:

1.   The product is not satisfactory.

2.   Because of 1, potential users don't use the product.

3.   Potential users ask for a better product.

4.   Because of 2, the request is denied.

So, because the product doesn't meet the needs of certain users, they aren't identified as potential users of a better product, they aren't consulted, and the product stays bad.

7.1.4 The product can create users

The Railroad Paradox suggests that the product can actually *create* a user constituency, where one never existed before. When the fifty-five mph speed limit was put into effect to save fuel, traffic fatalities fell as a side effect. People who previously didn't care one way or another about the speed limit now became interested in retaining it, in order to save lives.

Even bad systems create user constituencies who have an investment in keeping the system bad. When one-way tolls were proposed to replace two-way tolls for Manhattan Island, for instance, everybody thought it was the "perfect" design: Everyone is no worse off than before, and some people are better off. But the toll-takers' union and the garages holding the tow truck franchises didn't think it was a perfect design. Neither did the ice cream and newspaper vendors who made their living hawking to cars sitting in queues at the toll gates.

The economist Pareto first discussed the situation, now called Pareto Optimization, when a change left some people better off and no people worse off. The economist Veblen, however, observed that every change—whether considered wonderful or awful—adversely affected some people and benefited others. According to our experience, at least, Pareto was a dreamer, and Veblen a realist (see Figure 7-2).

Figure 7-2. Veblen's Principle: There's no change, no matter how awful, that won't benefit some people, and no change, no matter how good, that won't hurt some.

7.1.5 Are losers users?

Veblen's observations suggest that we think of users as not just those who will benefit from the product, but as anyone who will be affected. Who will lose from this approach? What will they lose? If we don't think of them in advance, they are likely to appear later and oppose the product, sometimes with great success. The institution of one-way tolls has often been delayed by toll-takers' unions, and the raising of speed limits to sixty-five mph on selected roads met serious and effective opposition.

We often hear of "user-friendly" products, but there may be user groups that you want to treat in an "unfriendly" way. For example, we want to make the tops of prescription bottles difficult for children to open—children are a "loser" group for that product for their own protection. Sometimes, we have to identify a loser group for someone else's protection. Medical records, personnel records, and credit records all need to be protected against inappropriate users. When building such information systems, we have to identify those loser groups and make specific requirements to ensure that the systems are "unfriendly" to them.

For some situations, we may even want to make the very *existence* of the product unknown to certain loser groups. This is certainly true of many security systems, which work best when potential penetrators are unaware that they exist. It's also true of some professional systems, like locksmith's tools, as well as products that we don't even know exist, because we ourselves are in the loser group.

7.2 A User-Inclusion Heuristic

Clearly, our job is not just *identifying* all users, but also deciding how they are to be dealt with—friendly, unfriendly, ignored, or whatever.

7.2.1 Listing possible user constituencies

In order to avoid too narrow a view of the user community, a possible strategy is to brainstorm a list of all possible users. Sometimes it is possible to list particular users by name—the president of a corporation, for example—but ordinarily we will name constituencies, or categories, of users. Here is a list of constituencies that emerged from one brainstorming session about the Elevator Information Device Project:

students	managers	professionals
children	handicapped people	computer programmers
venture capitalists	men	women
electricians	stockholders	shoppers
store owners	tourists	police
foreigners	members of congress	lawyers
homemakers	teachers	construction workers
mechanics	invading armies	terrorists
doctors	Russian spies	industrial spies
dogs	blind people with dogs	firefighters
loiterers	parents	old people
church people	janitors	surfers
gamblers	athletes	movers
stockbrokers	shoplifters	small children
working people	elevator operators	building maintenance

emergency rescue	muggers	delivery people
animal lovers	people with pets	equipment handlers
contagious people	frustrated people	impatient people
vandals	graffiti artists	graffiti haters
art lovers	communicators	passengers
claustrophobic people	neighbors	architects
building owners	elevator inspectors	cleaning people
power companies	liability insurers	elevator designers
information designers	prostitutes	elevator music composers
freight movers	blind people without dogs	deaf people
people in wheelchairs	non-English speakers	nonreaders
first-time riders	experienced riders	first-time users
smokers	strangers	lonely people

When we brainstorm such a user list as a separate activity early in the design process, we often think of highly innovative design features and functions. There are a number of "unthinkable" constituencies on the list. Just consider, for example, how many previously unimagined features and functions that "prostitutes" and "stockholders" suggest.

7.2.2 Pruning the user list

Although merely listing these potential users raises our consciousness of issues in the design of an Elevator Information Device, that's not all we can do with the potential user list. We've now defined almost a hundred possible user constituencies, yet we consider it axiomatic that no design could optimally satisfy a hundred different user constituencies. What are we to do?

First, some of these constituencies overlap. A single passenger could be a blind, male, impatient, art-loving stockbroker who happens to be in the building to shop for a pet because he is lonely.

Second, the categories are not independent with respect to the ultimate design solution. A design solution that emphasizes features for construction workers and athletes might trade away functions and features that would aid the handicapped, and vice versa.

In order to deal with such conflicts, designers must recognize and accept the old maxim, *There is no free lunch,* and consciously trade off one constituency against another. One way to do this is by applying a user-inclusion strategy.

The first step in the user-inclusion strategy is to list potential users as we have done. The second step is to assign each user constituency one of three values (F, I, U), according to the way the design is to treat them:

F   be very *Friendly* to them
I   *Ignore* them
U   be very *Unfriendly* to them

If we have 100 constituencies, there are 3^{100} different ways of assigning these letters, each representing a different choice of how users will be included in this design. This number of possible design targets is so large that it has 48 digits when fully written out, yet each focus could produce a somewhat different final solution.

To take one example, assume that all users are to be ignored except computer programmers, elevator designers, information designers, and composers of elevator music. This design might be an elegant tribute to our ability to produce art for art's sake, but might not serve the handicapped or satisfy the liability insurers. In any case, the solution would probably differ markedly from one that included features for terrorists, spies, muggers, loiterers, shoplifters, and vandals. In these cases, we might deliberately design user-*unfriendly* features.

The user-inclusion process can also focus our attention on otherwise acceptable people for whom we would want to make the system unfriendly. For instance, we might design features that prevent children or blind people from operating an elevator while the building is under construction. We can label such designs unfriendly, though perhaps a better label would be "paternalistic."

Of course, every designer defines user constituencies, but without specific user-inclusion activities, many of the choices are *implicit*. When implicit choices to include or exclude are made explicit late in the process, corrective action becomes expensive. The result is such products as appliances that cannot be operated by left-handed people, transports that cannot be used by people in wheelchairs, and buildings that are happy hunting grounds for muggers and rapists.

7.3 Participation

Not all users are created equal. Not all users will affect or be affected by the product in the same way. And not all users participate in the product development in the same way. To aid in planning for user participation, we find it helpful to classify participation on three dimensions (Figure 7-3):

1. Who: Is representation by surrogate, sample, or exhaustive?

2. When: Is participation continuous or only at discrete intervals?

3. How: Is information based on experience or experiment?

7.3.1 Who participates?

Ideally, every known or potential user should participate in the requirements process. For certain products, such as individual information systems or custom-built houses, such exhaustive participation is possible, and even the norm. For other products, however, exhaustive participation is at best exhausting, and at worst impoverishing.

One obvious way to save time and money is to *sample* the presumed population of users. Samples can be chosen in many ways. At the one extreme is a scientifically designed sampling procedure. At the other extreme, we just grab who-ever happens to be around. Usually, the best course lies somewhere in between, but the most important rules are these:

1.   Be aware that you are sampling.

2.   Be aware of your sampling method.

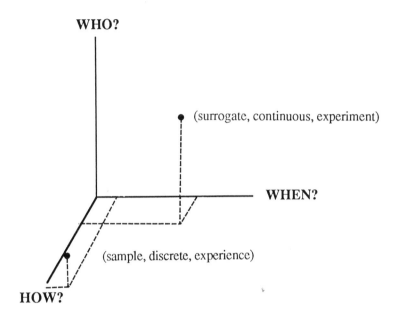

Figure 7-3. We can classify user participation on three dimensions.

For rule 1, the key phrase to listen for is, "We've talked to *everybody*." Talking to every user of a new product doesn't just happen. It takes lots of planning and dedication—more, in fact, than carrying out a reasonable sample. The United States Census Bureau claims that trying to count "everybody" costs millions more, and produces a less reliable census, than designing a careful sample. "We've talked to *everybody*" generally can be translated into, "We haven't given any thought to whom we talked to, but just talked to everybody *who happened to appear on the scene*."

Every way you seek users is a sampling method, and each different method includes a different group of users in a different way. For example, a group of ar-chitects placed a model of a company's new office building in the lobby of the present building. They placed questionnaires next to the model, to invite "everybody" to comment on the new design. But "everybody" who uses the building doesn't use the lobby, so this sampling technique excluded people who use the trade entrances, or who do not currently go to the lobby of this building, but will use the new building.

There's nothing wrong with using the model-in-the-lobby technique for sampling—indeed, it's a wonderful idea. What's wrong is believing that it's an exhaustive method of getting user information, rather than a *sampling* method. Because of this belief, the architects failed to notice the shortcomings of the sampling method actually in place, and thus failed to get information from several significant user groups.

Another common mistake is to use a *surrogate* for some part of the user population without realizing it. A surrogate is not a real user, but a stand-in. People from a marketing organization generally act as surrogates for a large population of potential buyers, a population that would be impractical to approach directly. Trouble occurs when the marketers forget they are only surrogates, and start to believe that they are the users. That's the beginning of all major marketing mistakes.

7.3.2 When do they participate?

The second dimension of user participation involves *time*. The users, or their surrogates, can be involved in the requirements process continuously, or at discrete intervals. In the one case, they're *on* the team; in the other, they're *on call* by the team.

We generally find users acting as full-time team members only in projects where the potential users are well known in advance, and are few in number. Even when users are on the team, rarely will *all* of them be included. You have to remember that your team member is a sample, not the entire user population. It can be somewhat embarrassing to ask for confirming opinions from other users not on the team, but it's essential that users on the team know that you will do so.

Part-time user participation is the far more common arrangement, but the major problem with such participation is that it easily drifts into zero-time participation. Users are interviewed once, then forgotten. Perhaps the reason is that the requirements team gets the impression the users don't want to be bothered, but that's seldom the case.

To overcome the difficulties of full-time and part-time participation, organizations often perform requirements work on a full-time basis, but only for a week or so at a time. After deciding who will represent the users, the team goes to a remote site and meets full-time for a few days or a few weeks, depending, of course, on the scope and complexity of the project. With this intensive effort, the major work of gathering user information is complete. What remains is testing, revising, and packaging the requirements.

7.3.3 How do we get their judgments?

Most of the time, users make judgments based on their perception of the requirements. As we've already seen, however, the user sometimes actually experiments with the requirements, perhaps using a mockup or a prototype. For instance, architects are now using computer simulations to walk people through buildings, shopping malls, or whole communities that exist only in their imagination—and as data stored in the computer model.

Once again, the most important thing is to realize whether your information is based on opinion or some actual trials. And, when it is based on trials, to what extent are the simulations realistic? When architects make scale models of buildings, the models emphasize some aspects and de-emphasize others. That's why so much expense goes into making the tops of buildings so ornate. New York's Chrysler building must have been an enticing model, but nobody on the street can really appreciate the beauty of its topmost decorations on the full-sized building.

7.4 Plan for Capturing Users

In any requirements work, all types of user involvement may be employed. We might use continuous judgment based on a marketing surrogate, supplemented by a monthly survey sent to a large sample of the user population, plus a discrete experiment with a small sample of users to calibrate the judgments. Rather than let all this happen willy-nilly, though, we need to create a user-inclusion strategy to consider just how each user—including all friendlies and unfriendlies—is to be handled.

Having a plan for user inclusion is not enough, however. We also need a commitment, as well as the ingenuity to execute it. It's one thing to decide which users we need, when, and for what kind of judgment; it's quite another to get them to agree to the involvement we want. In some cases, simply finding them is difficult.

With our clients, we always recommend some sort of *broadcasting* technique for capturing users. As the requirements work proceeds, we post an announcement in the company newsletter, on bulletin boards, via electronic mail, or whatever systems the organization uses to broadcast information. The announcement tells what we are working on, and solicits contributions. We may not get many responses, but the ones we get turn out to provide extremely valuable contributions to the requirements effort.

At one giant company, only one user responded to our request for input on a large purchasing system in the planning stage. It turned out, however, that this user was from the auditor's office, which had been overlooked in the planning. He proceeded to demonstrate why the basic approach assumed for the project was not auditable, and therefore not permissible. As a result, the project was scrapped. The team was disappointed, but not nearly as disappointed as if the system had been scrapped *after* it was built, at a projected cost of $20 million!

7.5 Helpful Hints and Variations

- Various consulting firms offer packaged services to facilitate short, intensive requirements sessions. Many of our clients have reported terrific experiences with their sessions.*

- When dealing with an unfriendly user population, you must decide whether or not to trust what they tell or show you. Conversely, if you use surrogates, can you trust the opinions of people who are not actually thieves and muggers? To get trustworthy information, some organizations actually hire the people who have breached their security systems. Rather than send them to jail, they use them as the most reliable experimenters they can find.

- To avoid dealing with criminals after the fact, some organizations sponsor legitimate competitions inviting people to try breaking through their security systems. Offering an attractive prize may adequately simulate the unfriendlies, and the results at least would be more trustworthy.

- A powerful surrogate method is the anthropologist's "participant observation." Members of the requirements team join the user population for a time, working right alongside them and making observations to bring back to the requirements process. This method is well worth trying.[†]

- Apply Veblen's abstract principle in the following very practical way: Whenever you hear someone say, "Nobody could possibly be hurt by this product," immediately initiate a "loser identification" brainstorm.

7.6 Summary

Why?
Adopt an explicit user-inclusion strategy and a plan to assure all users are dealt with consistently.

When?
Perform user-inclusion work before any design decisions are made, and the earlier the better.

How?
1. Agree in the requirements team that you must take charge of the user-inclusion strategy, rather than "just letting it happen."

2. Brainstorm a list of potential users.

---

*Five organizations offering this service with which we've had good experience are the following: Western Institute for Software Engineering, 11911 NE 1st St., Suite 206, Bellevue, WA 98005; SPS Sweden, Salviigrand 1, Box 2129, S-10314 Stockholm, Sweden; The Strategic Advantage, 3601 W. 29th St., Topeka, KS 66614; Swirczek & Assoc., 1710 Chaise Court, Carson City, NV 89703; CASEware Inc., P.O. Box 8669, Portland, OR 97207.

[†]For an excellent introduction to participant observation, one that requires no background in the social sciences, see James P. Spradley, *Participant Observation* (New York: Holt, Rinehart & Winston, 1980).

3.   Reduce the list by classifying them as friendly, unfriendly, and ones to ignore.

4.   Using the three dimensions of participation, develop a strategy for each user group you don't want to ignore.

5.   Carry out the inclusion plan, and use your imagination and tact to get the full participation you need.

Who?

Discovering the "who" is the whole purpose of a user-inclusion strategy.

8   MAKING MEETINGS WORK FOR EVERYBODY

There are two major types of adventure stories about exploration. In the first, the lone explorer faces the elements, showing ingenuity and problem-solving ability at every new crisis. In the second, a group of people face the elements together. They benefit from their multiple talents and points of view, and they pay for these benefits by the troubles they have in getting along together.

The lone explorer is the ideal image that many of us carry forward from our school days, when working cooperatively was called "cheating." Many organizations reinforce this negative image of cooperative work, encouraging "competition" among employees by such devices as individual achievement awards.

In any sizable development project, however, the idea of doing it alone is more of a fantasy than any Indiana Jones adventure. In order to obtain the capacity and diversity that a team can provide, we must give up our heroic fantasies and pay the costs of working with other people. In no place are these interpersonal costs more evident than in meetings. In this chapter, we'll explore the use of meetings as tools: how to make them more productive and less vexatious.

8.1 Meetings: Tools We Can't Live With, or Without

Meetings can be terrible. At some moment in our lives, every one of us has sworn we would never attend another meeting. We complain that they're too long and boring, they stray from the subject, they're painful, and they have no purpose. But then we always relent, because we can't live without meetings. Meetings are tools—social tools—and without them we could develop only the simplest products.

Before discussing meetings in general, though, let's look at a terrible meeting in painful detail to see what can go wrong.

8.1.1 A terrible, but typical, meeting

(Jack and Zara arrive at Room 1470B at 9:05 A.M. and find it totally dark. Jack turns on the light and looks at his watch.)

Jack: Where is everybody? The meeting was supposed to start five minutes ago.

Zara: More important, where's the coffee? And Danish? I had to skip breakfast to make this meeting. I'll go see if I can find out what happened to it.

Jack: Wait, don't go. . . .

(Jack sits down and starts going through his mail. At 9:12 Sid arrives, carrying a brown bag and a pile of mail. He takes out a cup of coffee and a large jelly donut, and sits down to start eating and reading without saying anything to Jack. At 9:17 Martha and Mary Jo arrive, also with coffee.)

Martha: Hello Jack! Hi Sid! Sorry we're late. Big line at the coffee trolley. Where is everybody?

Jack: Zara went to find out where the coffee and Danish are. Beats me where Ned and Retha are. . . . I didn't know you were coming to this meeting, Mary Jo.

Mary Jo: Well, I didn't either, but I met Martha in the coffee line and she said she thought I might be interested.

Jack: Did you interview any users?

Mary Jo: Well, no, since I didn't know I was coming.

Martha: That's okay, Mary Jo. I didn't get any interviews done either.

Jack: *(looks at his watch and grimaces)* It's 9:25, and I've got another meeting at 10:00. We'd better get started. The others can catch up when they get here. Sid, could you put away your mail?

Sid: Grhnhg.

Jack: What?

(Sid groans and closes his folder with a loud slap on the table. Jack goes to the whiteboard and picks up a green marker. He starts to write but the marker is dry. He tries the only other marker, a red one, and it's dry, too.)

Jack: Drat! Does anybody have a marker pen that works?

Martha: I know where some are. I'll go get them. *(She stands up and heads out the door, even as she's still talking.)*

Jack: Wait! Oh, all right, I suppose we can start without her. Sid, will you take notes?

Sid: Grnngh.

Mary Jo: Is there an agenda?

Jack: Ned was responsible for that, but he's not here. Anyway, the meeting is supposed to determine what our users think of the requirements document so far. Hey, here's Ned, at last!

(Ned and Zara come in, Ned carrying a large coffee dispenser and Zara carrying a round aluminum tray of Danish covered with plastic wrap. Jack looks conspicuously at his watch.)

Jack: Well, Ned, it's about time. It's 9:35.

Ned: I know. We're going to have to do something about the cafeteria. If Zara didn't go down there, they would have just let this coffee sit through our whole meeting. Fortunately, I met her as she was coming out, or she'd have had to make two trips.

Zara: Thanks, Ned. I appreciate it.

(Zara struggles to get the plastic wrap off the tray. Sid puts down his mail and comes over to help her. After a few minutes, they finally get one corner open, and Sid takes two cheese Danish and goes back to his seat to resume reading his mail.)

Zara: Come on, everybody! Get 'em while they're hot!

Jack: Could we get started? We're trying to make an agenda.

Ned: Why? I've got the agenda.

Jack: *(angrily)* Right! And you weren't here!

Ned: So, do you want to fight about it? I'm here now, aren't I?

Jack: *(cooling down)* No, let's get on with it. I was just telling Mary Jo that the agenda was to report on interviews with our users, about their level of satisfaction with the requirements so far.

Ned: Well, that's *your* idea, but the real purpose of this meeting is to find out if we've managed to reduce the ambiguity in the requirements, before we give it to the users.

Zara: You mean I wasn't supposed to give them the requirements document yet?

Ned: You silly fool! You mean you let them see *that* piece of junk? Can't you do anything right?

Zara: I thought that was the agenda. Didn't you, Sid?

Sid: Ggrnnhgn.

Zara: See!

Jack: Let's stop fighting, and get on with the meeting. So, we have *two* items on the agenda—user satisfaction and measuring ambiguity. Anything else?

Mary Jo: If there's time, I'd like to report on the status of the United Way drive. It won't take more than five minutes, or maybe ten.

Jack: *(tapping his watch)* We've only *got* twenty minutes. And I don't even have a good marker pen yet!

Mary Jo: I'm sorry, but I'm not responsible for the marker pen. Don't you think that United Way is a good cause? Think of all those poor children who get lunch supplements, and the old people who need someone to wheel them around, and take them out once in a while. Just because you're one of the privileged few, you don't have to be arrogant!

Jack: All right! All right! Take your ten minutes. Maybe everyone will be here by the time you finish.

(Mary Jo proceeds to report on the progress of the United Way drive. After seven minutes, Martha shows up with a handful of markers. After twelve minutes, Jack starts tapping on the table. After fifteen minutes, Sid packs up his mail, stands up, takes two more Danish, muttering something about "no more cheese," and leaves. At 10:00, the door opens and Millie sticks her head in.)

Millie: Are you guys almost finished? We have this room for a meeting at ten.

Jack: *(sighing)* Sure, we're finished. Come on in.

Millie: Thanks. Oh, boy, can we have what's left of your coffee and rolls?

Jack: Sure, help yourself.

(Everyone remaining gets up to leave.)

Millie: Hey, Martha, don't take those marker pens with you.

Martha: That's okay. There's a red one and a green one there on the board.

(Martha leaves, and Millie sits down with a raspberry Danish. She looks at her watch. . . .)

> Reader: As you read through this chapter, look back at the meeting of Jack, Zara, and the rest to see how many principles of productive meetings were violated, and what they might have done about them.

8.1.2 Meetings as measurements

Meetings like this are supposed to be tools for accomplishing work, but they often seem more like melodramas, displaying the entire range of human emotions. Meetings are social tools. Like other tools, they come in all shapes and sizes, depending on the job they are designed to accomplish. One job that *every* meeting can accomplish is to *measure the health of the requirements process.*

If your meetings are terrible, then your process is sick.

Noticing what's wrong with meetings will help diagnose just what's wrong with the entire process, and indicate what to prescribe for the malady.

8.2 Participation and Safety

Some of the most common things wrong with meetings are when people stomp out, fail to attend in the first place, or are there in body but don't seem to contribute. These actions are not just an annoyance, they are *symptoms of an unsafe climate.*

These symptoms indicate people are not participating fully in the requirements process. In order to get full participation, meetings must be made safe. How is this achieved?

People are different, and what will make one person feel safe can make another queasy. So to make the situation as comfortable as possible for everyone involved, *negotiate an agreement on ground rules before dealing with the content of the meeting.* As participants negotiate this agreement, specific reasons that people are not participating will become evident.

8.2.1 Establishing an interruption policy

For instance, many of us are in the habit of interrupting others, yet being interrupted makes some participants insecure about speaking up. Most interrupters, we find, agree that they don't want to interrupt, but they don't want to be interrupted, either. So we are usually able to get agreement up front that interruption isn't okay in our meeting. Also, nobody will object that this rule must apply to *everyone.* With that agreement in hand, the meeting facilitator can readily and politely silence interrupters.

8.2.2 Setting time limits

Agreement on time limits has a similar effect and, in fact, limiting individual speakers may be a precondition to an agreement on interruptions. Interrupters often worry that someone will simply keep the floor forever. If everyone agrees in advance that the facilitator can enforce, say, a two-minute time limit on each speaker, then the environment becomes safer for the interrupters, who can then more easily make it safer for everyone else.

8.2.3 Outlawing personal attacks and put-downs

Another important agreement that's easy to get up front is a prohibition on personal attacks or put-downs. Nobody will ever object to this rule in the calm beginning of a meeting, but many will violate it in the heat of events. Prior consensus on this rule gives the facilitator the group's permission to intervene in any personal attack situation and resolve it before it grows into a knock-down-drag-out battle (see Figure 8-1).

Figure 8-1. Agreeing in advance to refrain from personal attacks and put-downs will make it possible to resolve attack situations before they escalate.

8.2.4 Reducing pressure

Although some people thrive under pressure, a boiler-room type atmosphere shuts down many people altogether. Hence, every meeting needs *devices for staying out of emergency mode*. One important device is the *time-out*, when participants agree in advance that anyone can call one-minute or five-minute time-outs at any point without explaining why. They may need to get more information, to have time to think, to call home to get some crisis off their mind, or simply to use the toilet.

Agreeing that each participant is entitled to a reasonable number of unexplained personal time-outs, helps prevent meetings from being high-pressure events.

8.2.5 Allowing time to finish, yet finishing on time

Another pressure-reducing technique is to agree to allow all the time needed to get the work done right—but not necessarily today. In other words, stick to agreed time limits, but schedule a continuation of the meeting if business isn't finished.

One way to determine if there's more business is simply to ask each person, "Has every one of your ideas been handled to your satisfaction?" As one of our clients put it, "I like to be asked that question. I don't have to get my way every time, but when I feel I haven't been given a fair hearing, you can be sure I'll bring up my idea in other meetings until I feel I've been heard."

8.2.6 Handling related issues

What if their issues aren't really part of the business of this meeting? For those cases, use a *related issues list*. As the meeting begins, obtain agreement that issues irrelevant to this meeting will be posted on a large blank sheet of paper in full sight and labeled, "Related Issues."

When an issue comes up that seems out of order, the facilitator asks, "Is this meeting the proper place to handle your issue?" This question elevates the discussion from the details of the particular point and onto the level of the point's relevance at a specific time and place.

Participants will feel safe and satisfied if they know that their idea has been written down—and that the issues list is not just a fancy garbage can. To achieve this goal, assign each related issue to a person responsible for getting it to the proper people as soon as the meeting terminates.

8.2.7 Amending the rules

Whatever your meeting rules, always get agreement up front so the first transgressors won't think they're being picked on. Unfortunately, something always comes up that you hadn't anticipated. When that happens, handle the case fairly, then stop the process to get agreement on amendments to the initial rules.

For this reason, get advance agreement to amend the rules when something unexpected comes up. It also helps to agree in advance not to use the amendment process as a political tool, for one side to dominate another.

8.3 Making It Safe *Not* to Attend a Meeting

Meetings should be as small as possible—but no smaller. One of the common complaints about meetings is that *too many people attend*—people who really have little or no interest in the business being conducted. This might seem to be the simplest of all meeting problems to solve. After all, why in the world would people attend a meeting in which they have little or no interest?

The answer is that most will tolerate a great deal of boredom in order to buy safety. People don't want to miss a meeting that *might* be important to their work, so they go to a meeting where they probably don't belong *just in case something happens that might be important to them.*

In other words, the presence of unnecessary people in meetings is a measure of uncertainty over what's going on in the requirements process. In order to make it safe *not* to attend a meeting, use techniques to remove the *uncertainty* about what will actually be covered. Let's look at four such techniques.

8.3.1 Publishing an agenda and sticking to it

The first safety technique is to publish an agenda well in advance and not to deviate from it, no matter how tempting. By deviating, you may handle the side issue well, but doing so sends a dangerous message to all those excluded from the meeting that they should have attended to protect their interests (Figure 8-2). Next time, they'll ignore the agenda and show up, just to be safe.

8.3.2 Staying out of emergency mode

If you don't want to punish people who believe the agenda, you need a way of handling emergency issues that doesn't hurt people who don't attend the meeting.

The first principle, of course, is to *stay out of emergency mode.* Rarely in requirements work is there a true emergency issue that can't be postponed until it can be put on a published agenda. If you habitually find yourself yielding to the emergency argument, then perhaps you should pay attention to the real cause of this symptom. A project that's in emergency mode during requirements work will be in death mode when it comes time to deliver a product.

8.3.3 Handling people who don't belong

Even when you publish agendas and stick to them, there still will be unnecessary people in attendance. They may not have received a proper hearing in other meetings, so they attend whatever meeting is available, hoping to find a willing audience. To prevent this problem, be sure that participants don't leave meetings unheard (see Section 8.2.5).

If surplus people show up anyway, don't pretend they're not there and hope they won't interfere. Confront them immediately and if possible unobtrusively. The

Requirements Meeting, 10/27, 8:00am - 8:50am.

AGENDA

Develop list of flange attributes.
    (15 min.)

Develop list of surface finish attributes.
    (15 min.)

Develop list of retarder-spring attributes.
    (15 min.)

Create agenda for next meeting.
    (5 min.)

Figure 8-2. Posting an accurate agenda in advance—and sticking to it—will help keep out people who don't belong in the meeting.

easiest time for a surplus person to leave and still save face is before the meeting officially starts. Once people have been there a while, it's hard for them to admit they were wrong. Instead, they are likely to start participating to "prove" they were right to be there in the first place.

8.3.4 Including the right people

Another reason people show up at meetings where they obviously don't belong is that they *weren't* invited to meetings where they *do* belong. The first rule of successful meetings is to be sure the right people are involved, as we discussed in Chapter 7.

8.4 Designing the Meeting You Need

Applying these general rules to every type of meeting assures a safe and healthy project culture. However, even with this culture of safety as a background, you'll still have difficulty with meetings unless you learn to tailor each meeting to the job at hand. Designing each specific type of meeting so it has its own appropriate climate will assure not just safety, but also productivity.

For example, there are meetings to disseminate information, to gather information, to raise everybody's spirit, to increase the number of ideas, and to decrease the number of ideas—each with its own structure, pace, and rhythm. A meeting that tries to do two different types of job winds up doing neither job well. Even if the structure is right, the meeting will be larger if it covers two distinct topics, as illustrated in Figure 8-3.

The first step in tailoring a meeting is to decide on its single purpose. If you need a pep rally and a technical review, for example, make two meetings. If they are scheduled one right after the other with the same people in attendance, take a break and change rooms. At the very least, change seats.

The final rule of successful meetings is the same as the Boy Scout Motto: *Be prepared.* Ninety-five percent of all meetings that fail could have been predicted as failures before they ever started because of inadequate preparation. A checklist can take care of all the administrative matters, such as having fresh marker pens. To be sure every checklist makes sense, start it with the number one preparation question:

Do you have the appropriate design for this type of meeting?

8.5 Helpful Hints and Variations

- As a guide to the design of meetings of all sorts, we have used *How to Make Meetings Work** with great success. It's an inexpensive paperback, and we often provide a personal copy to every participant in a project. It's always paid for itself in one meeting.

---

*Michael Doyle and David Straus, *How to Make Meetings Work: The New Interaction Method* (Chicago: Playboy Press, 1976).

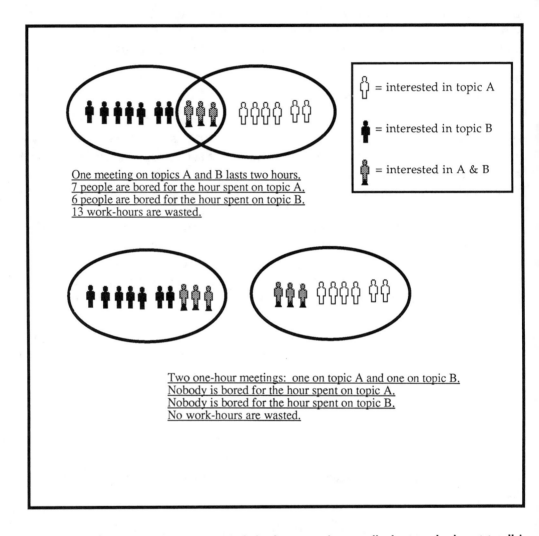

Figure 8-3. Keeping meetings to one topic helps keep meetings small, short, and relevant to all in attendance.

- If your project seems to hold too many meetings, that may be a symptom of overstaffing. Alternatively, it may mean that the project organization is not broken down along natural lines of cleavage, so that nobody can do anything without affecting everyone else.

- Within any meeting, you can learn enormous amounts about various aspects of the project by *observing individual behavior*. We'll consider this use of meetings in Chapter 13 on tools for facilitating meetings.

- Rescheduled meetings promote a ripple effect, leading to more rescheduled meetings, which in turn lead to more rescheduled meetings, and so on. If meetings are frequently rescheduled, or canceled, this may be a symptom of a lack of planning, or overload, or simply a project completely out of control.

8.6 Summary

Why?
Because of their central role in exploring requirements, meetings must be considered like any other tool: Design them, select them appropriately, train everyone in their use, and practice, practice, practice. In particular, use them as devices for measuring the health of the project environment.

When?
Meetings are used all the time. They may be formal, scheduled meetings, or informal meetings held in the corridor.

How?
Keep in mind the following:

1. Create a culture of safety for all participants.

2. Keep each meeting as small as possible, but no smaller.

3. Limit each meeting to a single type, and design that type to the job at hand.

4. Be prepared.

5. Use skilled facilitators, a subject to be discussed in Chapter 13.

Who?
Meetings involve everyone. Specific meetings may include fifty people, or two. We never know when we might find ourselves in a meeting, so we must always be prepared. Choosing the right people to be in the meeting, and those to exclude, is one of the most important parts of being prepared.

9   REDUCING AMBIGUITY
FROM START TO FINISH

The fundamental problem of requirements definition is ambiguity. As we saw in Chapter 3, the "How many points on the star?" question simulates the presence of ambiguity in real requirements work. As you'll recall, not one of the hundred participants was able to state the original problem statement exactly as it had been posed, yet most of the remembered questions started with the words, "How many points." It's as if everyone listened to the first few words and then filled in what they each thought the problem statement was going to be. This simulation indicates that if we are ever to reduce problem statement ambiguity we'll need tools to slow down the mind, tools to prevent us from racing ahead to try solving a problem prematurely. In this chapter, we'll develop several such tools or heuristics to identify sources of ambiguity in the problem statement.

9.1 Using the Memorization Heuristic

Many psychological experiments have demonstrated that *meaningful information is easier to recall than meaningless information.* For instance, computer programmers can find errors in code by trying to memorize code sections and then recall them from memory. The parts that are hardest to remember are those that don't make sense, and are in fact more likely to contain errors. The same is true of parts of requirements documents.

The "How many points" part of the star question may have been recalled by everyone not just because it was the first part of the question, but because its meaning was clear. We can use this relationship between memory and meaning to construct a heuristic for identifying problem statement ambiguity.

To apply this heuristic, ask individuals working on the problem to recall a specific problem statement precisely from memory, just as in the star simulation. Parts that are not remembered well by the group are likely to be places where meaning is not clear, and therefore rich sources of problem statement ambiguity.

A variation of the method can be used even when there is no explicit problem statement. Ask pertinent individuals to list what they consider to be the critical events and the critical aspects of the desired problem statement. Naturally, they should do this task as independently as possible, so as to maximize differences when the lists are compared and discussed.

After a period of discussion, the same individuals again independently define, as explicitly and completely as they can, the same events and critical problem aspects. Once again, compare the definitions and note differences. Then use an ambiguity poll to measure progress and decide whether more work is needed, and where.

9.2 Extending the Ambiguity Poll

The ambiguity poll is another useful tool to reveal sources of problem statement ambiguity. The poll itself measures ambiguity at one moment in time, but successive polls can provide information about shifts in meaning. For instance, in our seminar, when we revealed the exact original problem statement, "How many points were in the star that was used as a focus slide for this presentation?" some participants were amazed at their inability to recall certain specific words in the statement. Following discussion of these issues, participants were asked to estimate the correct answer for a third time, giving the distribution in Figure 9-1.

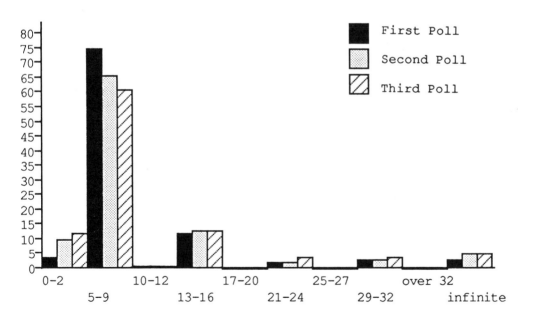

Figure 9-1. Shifts in answers after participants saw the problem statement.

Only a few of the hundred individuals changed their minds (or else there were compensating changes), so we might be tempted to conclude that *problem statement ambiguity* was only a minor component after all. Were the participants intuitively correct in not recognizing it in the first place?

Before drawing this hasty conclusion, we looked for possible trends or patterns among those who *did* change their minds upon seeing the problem statement. By focusing on understanding these changers, rather than trying to force them to accept the majority view, we discovered new ambiguous aspects to the problem.

9.3 "Mary had a little lamb" Heuristic

In real life, however, we seldom have a hundred participants to poll, so we need additional heuristics for identifying problem statement ambiguity. In addition to the memorization heuristic and the ambiguity poll, we need a tool that will readily identify ambiguity of problem statements in written form, since problem statements are normally transmitted that way from client to designer, once they are formalized.

Nursery rhymes are infamous examples of ambiguity because some of them have been transmitted from child to child over hundreds of years, and the original meaning of the rhyme may have been lost, or transformed. For instance, "Pop goes the weasel" is not about an exploding animal, but about pawning one of a shoemaker's tools, the anvil, which was called a "weasel," perhaps because its shape resembled that animal. To "pop" meant to pawn, hence the rhyme:

> A penny for a spool of thread,
>   A penny for a needle,
> That's the way the money goes,
>   Pop goes the weasel.

Now the puzzling rhyme makes more sense: The shoemaker has a cash flow problem because of rising expenses for small items, and so pawns his anvil to tide him over.

It's not surprising, then, that we base our next powerful heuristic on another familiar nursery rhyme:

> Mary had a little lamb.
>   Its fleece was white as snow.
> And everywhere that Mary went,
>   The lamb was sure to go.

Is it possible that this rhyme, too, contains hidden meanings?

If we *emphasize each of the words in the line, one by one, and then in combinations,* we can easily identify 6, and with a little more effort, 32 different meanings of just the first line of the familiar poem. To illustrate:

> *Mary* had a little lamb.
> (It was Mary's lamb, not Tom's, Dick's, or Harry's.)

Mary *had* a little lamb.
(She no longer has the lamb.)

Mary had *a* little lamb.
(She had only one lamb, not several.)

Mary had a *little* lamb.
(It really was surprisingly small.)

Mary had a little *lamb.*
(She didn't have a dog, cat, cow, goat, or parakeet.)

Mary had a little lamb.
(John still has his little lamb.)

As a tour de force, we offer all five words emphasized:

Mary had a little lamb.
(As contrasted with Pallas, who still has four large turtles.)

The remaining combinations you can work out for yourself, as an exercise. When you're done, you may realize that the most difficult of all the lines to understand is the original, unemphasized, version:

Mary had a little lamb.
(Why are you telling us this?)

Although the nursery rhyme example may seem silly, we've used this heuristic for years in reviewing requirements documents, and it has often saved vast sums of money.*

9.4 Developing the ''Mary conned the trader'' Heuristic

Another valuable heuristic for reducing problem statement ambiguity can be illustrated with the same nursery rhyme. In this heuristic, we *substitute synonyms for each of the key words* taken out of our context. After all, the client may be producing the problem statement in another context, which would be a major source of ambiguity.

---

*We first explained this technique in our little book on problem definition, *Are Your Lights On? How to Know What the Problem Really Is,* 2nd ed. (New York: Dorset House Publishing, 1989). See also "Functional Specifications Review" in Daniel P. Freedman and Gerald M. Weinberg's *Handbook of Walkthroughs, Inspections, and Technical Reviews,* 3rd ed. (Chicago: Scott, Foresman and Co., 1983) for more technical examples of this technique of stressing words, plus variations, and other heuristics for inspecting requirements documents.

Examining just the two words "had" and "lamb" provides a few surprising interpretations. Here are some alternate dictionary definitions that are particularly rich in possibilities:*

had -		past of have
have -	1a:	to hold in possession as property
	4a:	to acquire or get possession of: OBTAIN (best to be had)
	c:	ACCEPT: to have in marriage
	5a:	to be marked or characterized by (have red hair)
	10a:	to hold in a position of disadvantage or certain defeat (we have him now)
	b:	TRICK, FOOL (been had by a partner)
	12:	BEGET, BEAR (have a baby)
	13:	to partake of (have dinner)
	14:	BRIBE, SUBORN (can be had for a price)
lamb -	1a:	a young sheep esp. less than one year old or without permanent teeth
	b:	the young of various other animals (as smaller antelopes)
	2a:	a person as gentle or weak as a lamb
	b:	DEAR, PET
	c:	a person easily cheated or deceived esp. in trading securities
	3a:	the flesh of lamb used as food

By combining these definitions in various ways, we can discover many new interpretations of "Mary had a little lamb."

have	lamb	Interpretation
1a	1a	Mary owned a little sheep under one year of age or without permanent teeth.
4a	1a	Mary acquired a little sheep under one year of age or without permanent teeth.
5a	1a	Mary is the person who owned a little sheep under one year of age or without permanent teeth.
10a	1a	Mary held a little sheep under one year of age or without permanent teeth in a position of disadvantage or certain defeat.

*We adapt from *Webster's Seventh New Collegiate Dictionary* (Springfield, Mass.: G. & C. Merriam Co., 1967). Use of the *Oxford English Dictionary* would produce even more fascinating results.

10b	1a	Mary tricked a little sheep under one year of age or without permanent teeth.
1a	1b	Mary owned a young antelope.
12	2a	Mary is (or was) the mother of a particular small, gentle, person.
13	3a	Mary ate a little of the flesh of lamb.
14	2c	Mary bribed a small person trading in securities who was easily cheated.

Other interpretations can be developed as an exercise in the method. (We've left out some of the more racy definitions, to protect the lambs.) By examining the other lines of the poem, we can help sort out the true meaning. For instance, the second line says,

Its fleece was white as snow.

Examining "fleece" and "snow" in our dictionary, we find

fleece –   v,2:   to strip of money or property by fraud or extortion
snow –   v,2b: *slang:* to deceive, persuade, or charm glibly

We can now see that the poem was actually about a charmingly glib but fraudulent person named Mary who tricked a small, helpless, gullible stock trader and stripped him of all worldly goods through deceit and extortion. Is it any wonder that

Everywhere that Mary went,
   The lamb was sure to go.

What else could the poor lamb do after she fleeced him?

In short, the poem is making an editorial comment on today's business climate. The commonly held view that the poem is about an innocent young lady with a loyal pet is naive in the extreme—not as naive, though, as sophisticated adults who pick some line out of a requirements document and without giving it a second thought, proceed to develop a product based on a single, wrong, interpretation.

9.5 Applying the Heuristics to the Star Problem

To see how easily such naive product developers could have saved themselves millions, let's apply these two heuristics to the star problem. It's a bit closer to real life than a nursery rhyme, and so will illustrate the value of the heuristics more directly. We will especially want to see if new, plausible interpretations of the problem emerge.

Once again, the problem statement is

How many points were in the star that was used as a focus slide for this presentation?

First we apply the "Mary had a little lamb" heuristic:

How many points. . .
How *many* points. . .

do not seem to provide any new insight. But when we try

How many *points*. . .

we immediately see new possibilities. The real problem statement ambiguity may be in the word "points," so we consult our dictionary:

point –	1a(1):		an individual detail: ITEM
	(2):		a distinguishing detail
	b	:	the most important essential in a discussion or matter (point of a joke)
	c	:	COGENCY: PURPOSE
	4a(1):		an undefined geometric element of which it is postulated that at least two exist and that two suffice to determine a line
	(2):		a geometrical element determined by an ordered set of coordinates
	b(1):		a narrowly localized place having a precisely indicated position
	(2):		a particular place: LOCALITY
	c(1):		an exact moment
	(2):		a time interval immediately before something indicated: VERGE (at the point of death)
	d(1):		a particular step in development (boiling point)
	5a	:	the terminal usually sharp or narrowly rounded part of something: TIP
	6a	:	a projecting usually tapering piece of land or a sharp prominence
	b(1):		the tip of a projecting body part
	c(1):		a railroad switch
	(2):		the head of the bow of a stringed instrument
	7	:	a short musical phrase esp. a phrase in contrapuntal music
	8a	:	a very small mark
	b(1):		a punctuation mark; esp.: PERIOD
	(2):		DECIMAL POINT
	9	:	a lace for tying parts of a garment together used especially in the 16th and 17th centuries

10	:	one of nine divisions of a heraldic shield or escutcheon that determines the position of a charge
11a	:	one of the 32 equidistant spots of a compass card
14	:	one of the 12 spaces marked off on each side of a back-gammon board
15	:	a unit of measurement: as
a(1):		a unit of counting in the scoring of a game or contest
(2):		a unit used in evaluating the strength of a bridge hand
b	:	a unit of academic credit
c	:	a unit used in quoting the prices of stocks, shares, or various commodities
d	:	a unit of about 1/72 inch used esp. to measure the belly-to-back dimension of printing type
16	:	the action of pointing: as
a	:	the rigidly intent attitude of a hunting dog marking game for a hunter
b	:	the action in dancing of extending one leg so that only the tips of the toes touch the floor
c	:	a thrust or lunge in fencing
17	:	the position of a player in various games (as lacrosse)

We've probably made our point; otherwise we could point out some additional definitions and print them in 48-point type so as to point you to the right point of the compass. But that probably wouldn't score many points with a busy reader, since even with this expurgated list, the situation appears hopeless at this point. Let's try anyway.

In the context of Don's star, the definition in 5a appears to be the one most expected. By this interpretation, the question would read:

How many *terminal sharp or narrowly rounded parts* were in the star that was used as a focus slide for this presentation?

But there are a few other rather plausible interpretations suggested by the dictionary. In the context of the slide as an audiovisual device, definition 1c makes a lot of sense:

How many *purposes* were in the star that was used as a focus slide for this presentation?

In other words, how many reasons were there for showing the star slide? We could imagine more than one such purpose to the star slide, such as,

1.   to focus the slide projector

2.   to see if anyone was really paying attention

3.   to adjust the zoom to frame the following slides properly

4.   to call the seminar to order

5.   to introduce the topic of ambiguity and its many sources

How many *end points of straight lines* were in the star that was used as a focus slide for this presentation?

In this interpretation, we would count the internal points of the star as well as the external ones, giving 14 as the answer for a 7-pointed star.

How many *places or localities* were in the star that was used as a focus slide for this presentation?

This is what we would expect to see in a map, but was there any way of interpreting this slide as a map?

How many *exact moments* were in the star that was used as a focus slide for this presentation?

There was the moment at which the speaker began to speak. Then there was the moment at which the speaker mentioned that he likes "to use this slide as a focus slide." Or the moment when the slide vanished from the screen, or the moment it was mentioned again.

How many *very small marks* were in the star that was used as a focus slide for this presentation?

Some people recalled seeing dots, or "very small marks," inside the star and thus based their estimates on this interpretation. In a discussion with the participants, we discovered that this explained the cluster at one or two points, and also that the geometric definitions resulted in the cluster at infinity.

Other words in the original question also give interesting interpretations:

How many points were *in* the star that was used as a focus slide for this presentation?

Emphasizing "in" lends credence to those who interpreted a point as a dot or spot.

How many points were in the star that was *used* as a focus slide for this presentation?

Which star was actually used to focus the projector? Don said, "I like to use this slide (meaning the first) as a focus slide for this presentation," but he didn't actually focus the projector until the next slide was on the screen. The second slide, called "Convergent Design Processes," had many stars depicting *points* on two curves. Some people would call them asterisks, but many computer professionals call them stars. Programmers refer to the combination "**" as "star dot star."

How many points are there in each of these stars? Could be eight if we count the ends, or one if we consider the intersection of the lines as a point. In the ensuing discussion, we discovered that some participants considered the second slide to be the focus slide and counted one point per star to produce the cluster at 29-32.

That cluster also included the response of someone who remembered the question as,

> How many *stars* were in the *slide* that was used as a focus slide for this presentation?

This interpretation surprised even Don, who remembers this case as a striking demonstration of the keen powers of observational, recall, and problem interpretation—combined with a little mistake in memory leading to an ingenious solution to the wrong problem.

But we're not quite finished:

> How many points were in the star that was used as a focus slide for *this* presentation?

Which presentation? the entire seminar? the part that commenced after the coffee break with this now-famous question? In the second case, we must consider the first slide following the coffee break, which asked,

> **How many points were in the star that was used as a focus slide for this presentation?**

There's *one* "points" (the word) on this slide. The type used was 12 points in size. The discussion it engendered made dozens of points. But, was this slide really used as a focus slide? It certainly was, in the sense of focusing the attention of the seminar on heuristics for reducing problem statement ambiguity.

9.6 Helpful Hints and Variations

• If people take a "Who cares?" attitude, none of these tools will be successful. It's easy, because of the sometimes ridiculous results, to believe that applying the "Mary had a little lamb" heuristic and the others is a silly game. Watch for that attitude, and confront it by asking people to be patient with the technique. Once they uncover their first serious ambiguity, they'll no longer need convincing.

- On the other hand, these heuristics require a playful atmosphere to be fully successful. If things get *too* serious, they're not working well because, by their nature, they'll always produce some ridiculous interpretations. If not, the atmosphere is not right, so get the group to lighten up.

9.7 Summary

Why?
The fundamental problem of requirements definition is ambiguity, and these heuristics are powerful tools for reducing that ambiguity.

When?
Use the various heuristics to uncover sources of ambiguity due to errors in

observation:	variations in what people *see* or *hear* of a critical event (intra-cluster variation)
recall:	variations in what people *remember* of a critical event (intra-cluster variation)
interpretation:	variations in how people *interpret* important happenings associated with a critical event (inter-cluster variation)
problem understanding:	variations in how people *define* in their own mind the problem to be solved (inter-cluster variation)

How?
In summary, there are four heuristics for uncovering ambiguity:

1. *the ambiguity poll:* Create a metric (such as performance, time to complete, or cost) that requires a solid understanding of the problem to estimate. Give the problem statement (requirements) to informed individuals and ask them to independently estimate the value of this metric. Bring the individuals together to compare and discuss the tabulated results.

2. *the memorization heuristic:* Various individuals try to recall the problem statement precisely from memory. Parts that are not remembered well are likely to be places where meaning is not clear.

    A variation of the memory method is to ask individuals what they consider to be the critical parts of the problem statement. When the lists are combined and discussed, differences reveal potentially ambiguous parts.

3. *the ''Mary had a little lamb'' heuristic:* Read the statement in question aloud several times, each time emphasizing a different word or words until as many interpretations as possible are discovered.

4. *the ''Mary conned the trader'' heuristic:* Determine the key operational words in the statement and list all their definitions. Mix and match these definitions to form additional interpretations.

Who?

Teach these heuristics to everyone who will ever have to construct or interpret a requirements document.

PART III   EXPLORING THE POSSIBILITIES

Human beings are not only explorers, they are also tool users and tool makers. Some of these tools are *physical,* such as chisels and computers. Others are *non-physical,* like chiseling techniques and computer programs. The word "tools" is sometimes used in the narrow sense to mean just physical tools, but in this book we use it in the wider sense to include techniques and programs.

Whenever people explore, they augment their natural capacities with various tools. They have tools for *moving,* such as legs, llamas, canoes, Jeeps, and helicopters; and tools for *looking,* such as eyes, ears, noses, professional scouts, telescopes, microscopes, radar, and spy satellites. Tools exist for *recording,* such as memory, maps, notebooks, and blazes cut into trees. And there are tools for *analyzing* and *deciding* which direction to go next, such as intuition, human guides, compasses, watches, guide books, and maps.

This book is filled with requirements tools that are similar in function, though quite dissimilar in appearance, to ordinary exploration tools for moving, looking, recording, analyzing, and deciding. Most of them are independent, general purpose, and portable, which means they can be carried in the head, which is appropriate for exploring *ideas.* They can be used alone, in combination, or with any additional tools commonly used for exploring requirements.

Figure III-1 represents a model of this exploration as a journey through the "Jungle of Ideas." When the helicopter lands in the jungle, it seldom lands at the ultimate destination, which is the "Golden Temple of Just the Right Ideas." Mostly, it lands in the North—too many ideas—or in the South—not enough ideas. It's up to the explorers to use their tools to find the Golden Temple.

In Figure III-2, we see the "Spiral of Exploration" by which explorations gradually converge on the Golden Temple. Explorers have tools to tell them whether they are north or south of the Temple, and other tools to transport them in the right direction. As you read about the tools in this book, notice how they fall into three categories: tools to tell you and record where you are (too many ideas, too few ideas, or Just the Right Ideas); tools to move you south (by reducing the number of ideas); and tools to move you north (by increasing the number of ideas).

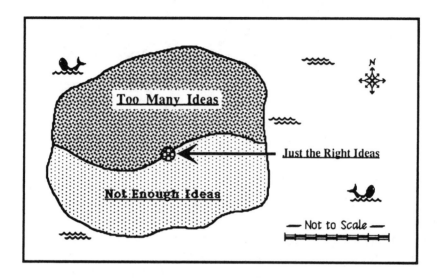

Figure III-1. The Jungle of Ideas hides the Golden Temple of Just the Right Ideas.

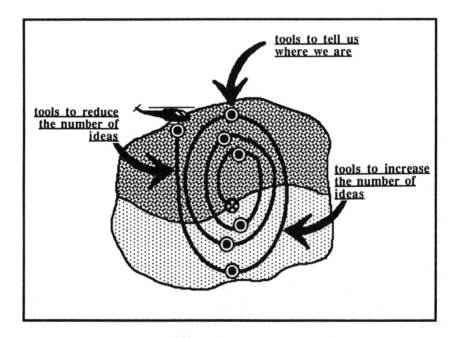

Figure III-2. The Spiral of Exploration suggests that the process of developing requirements is not a linear one, but cycles around its goal, ideally getting nearer and nearer.

The following chapters will discuss exploration tools for discovering some of the exciting destinations that are possible in the requirements process.

10   IDEA-GENERATION MEETINGS

It takes more than good intentions to create a meeting in which ideas can be generated. It also takes more than good general meeting rules. What it takes is a meeting *designed* for idea generation. The archetype of such meetings is the brainstorm, a meeting format designed by Alex F. Osborne, but much adapted over the past forty years. Many of our heuristics use brainstorming as an essential component.

Brainstorming is a term that has passed into the common language, with many different interpretations. In our experience, when people say, "Let's brainstorm!" the result is often "brainblizzards" instead—they freeze your brains, bury you under mounds of snow, and leave you cold. Let's look at a typical brainblizzard, to set the stage for better ways to generate ideas.

10.1 A Typical Brainblizzard

Jack, Zara, Martha, Sid, Ned, and Retha arrive at Room 1470B at 8:50 A.M. There is a carafe of coffee and a huge pile of jelly donuts, which they dig into with relish. At 9:00, Jack steps to the front of the room, picks up one of several marker pens, and writes the agenda on the board: IDEAS ABOUT REQUIREMENTS FOR THE DO NOT DISTURB PROJECT.

Jack: Okay, thanks to everybody for being here, and on time. We've got new markers and nice refreshments, so it should be a great meeting. Who's got an idea?

Zara: What's the Do Not Disturb Project?

Jack: You should know we're working on a way to get better traveler acceptance of our client's hotel.

Zara: Well, I call that the Happy Traveler Project.

Martha: Oh, you mean the Ritz Hotel Project?

Jack: It doesn't matter what we call it, it's client X's project. Let's get some ideas down on paper.

Martha: Okay, I think that we need to give the impression that this is a deluxe-class hotel.

(Jack makes a sour face and writes something on his notepad.)

Sid: Polished brass.

Jack: What are you talking about?

Sid: Elegance.

Martha: I think he means that polished brass will give an impression of elegance.

Retha: Polished brass is too expensive.

Ned: Not really. Not compared with some other materials. I mean, we could have gold, or at least gold plate.

Retha: Yes, gold costs more to start with, but brass has to be polished every day, or else it gets grungy. So you have an ongoing labor cost that has to be factored in with the initial acquisition cost.

Zara: Well, I like gold! It's a nice touch.

Sid: Robbers.

Jack: What does *that* mean?

Martha: I think he means that people will steal gold fixtures.

Retha: Well, that has to go into the cost equation, too.

(Retha stands up, goes to the board, and starts to write an equation. Sid gets up, grabs another marker, and sketches what looks like handles for a shower.)

Jack: Could we get back on track? We're trying to make a list of ideas for the Do Not Disturb Project. I'd like to see what the Do Not Disturb sign will look like.

Ned: Why? Until we have designs for the handles, there isn't anything to hang the sign on.

Sid: Hooks.

Jack: I'm going to ignore that. What about the sign?

Ned: I think we're getting into too much detail at this stage. After all, we're not trying to design the sign, but only to give requirements. So, we have to design the handle first, or we don't know if the sign will stay on the handle. I was in a hotel last week that had these lever kind of handles, you know, the kind that looks like those European window latches. Well, whenever I hung out the sign, then I'd turn the handle to close the door and the sign would fall off. So while I was in the shower, the maid came in. What a scene, me try-

ing to grab a towel, and she didn't speak any English, so she was
trying to tell me something and I couldn't make out a word. . . .

Jack: We're getting off track. Can you make it shorter?

Ned: Well, *you* had plenty of time to give *your* ideas. And I don't see you
        writing down *anything* I said.

Zara: You know, we should have one of those automatic translating com-
        puters in every room, so you could communicate with the maids.

Ned: That's a stupid idea. The maids can't read; if they could, they
        wouldn't be working as maids in a hotel.

Zara: Then we could get one of those machines that talks.

(Jack looks at his watch, then taps on his notepad.)

Jack: Well, I think I've got enough ideas now, and we're out of time any-
        way. Remember, we promised ourselves we'd end our meetings on
        time. Thanks for all the great ideas. I just know that our customer's
        going to love this Do Not Disturb Project. Thanks again!

Reader: As you explore the rules for brainstorming, notice how many
of them Jack and his colleagues violated in their abortive attempt to generate fresh ideas for the Do Not Disturb Project.

10.2 First Part of the Brainstorm

Osborne described a specific method, but you don't have to stick slavishly to
Osborne's exact format. You can use his four essential rules as base elements of
the design, and embellish each one to suit the idea generation you have in mind.

10.2.1 Do not allow criticism or debate

Osborne insisted that both adverse and positive judgment of ideas be withheld
until later. As ideas are generated, the recorder writes down every one of them,
like an automaton. Why? This ensures that ideas are not lost, people do not cen-
sor themselves, and the meeting doesn't go off on tangents with participants arguing over just how good or foolish an idea really is. Most of the world's great ideas
seemed "foolish" when they were first enunciated.

Of course, most foolish sounding ideas really do turn out to be foolish, and
some people don't feel safe if "foolishness" is written down. For this reason, the
unedited list of ideas should never be seen outside of the meeting. The agenda
must contain time for some editorial process to be applied to the list, once the brainstorm is finished. If participants understand and believe this promise, they can

relax and generate many "foolish" ideas, only one of which has to be a diamond in the rough.

Figure 10-1.   Do not allow criticism or debate.

10.2.2 Let your imagination soar

Osborne said that the wilder the idea, the better. Hence, the meeting environment must be safe so that people aren't afraid of looking silly. How can this be done? The facilitators can get things started by offering a few "silly" ideas. They can pass out funny hats, which seem to inspire participants to assume different personalities. Most important, facilitators must prevent any commentary that is overly serious or at worst derisive. If nobody's laughing, it's not a good session (Figure 10-2).

10.2.3 Shoot for quantity

Osborne noted that more ideas means more rough diamonds, and consequently the brainstorm must be designed to produce a quantity of ideas. For instance, pressure to "not waste time" is an insidious form of censorship. If wasting time is of genuine concern, set a time limit and challenge the group to produce some ridiculously large number of ideas within that limit.

Figure 10-2. Let your imagination soar.

Use all the time allotted, even if things bog down. Facilitators may be tempted to provoke more ideas, but instead, they should *be patient with silence.* Usually a long silence comes just before a breakthrough idea (Figure 10-3).

Figure 10-3. Shoot for quantity.

10.2.4 Mutate and combine ideas

In proper brainstorm design, participants are encouraged to suggest variations on
ideas already listed, or to combine ideas to create still more. These "improvements"

are not to be taken as criticisms of the original idea, but only as variations. Since all variations are new ideas, they are written down.

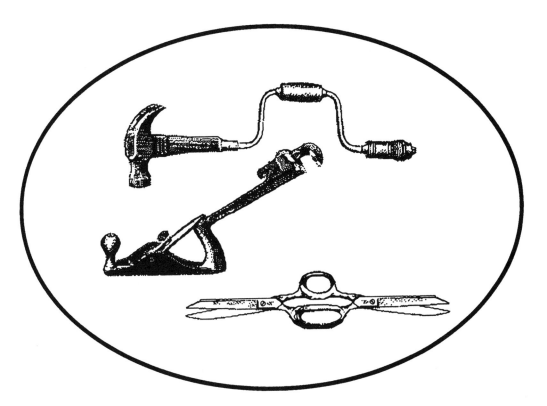

Figure 10-4. Mutate and combine ideas.

To facilitate the variation process, we like to have all ideas written in full view, on large sheets of newsprint, so they will be visible at all times. If necessary, use two or more recorders and have other people assigned to tear off full sheets and tape them to the wall. If the facilitator also acts as recorder, the meeting may suffer from too much control, which discourages ideas. Choose a recording method that will maximize the meeting's energy, and change the method if the energy is squelched.

10.3 Second Part of the Brainstorm

The first part of the brainstorm is designed to increase the number of ideas, and the second part is to reduce the list of ideas to a workable size. There are many potential idea-reduction methods, which can be used singly or in combination according to need. Five such methods are discussed below.

10.3.1 Voting with a threshold

In threshold voting, all participants receive a quota of, say, five votes each, which they individually tally by writing directly on the posted sheets of newsprint. Participants can give more than one vote to an idea, but they each get only five votes. Any participant can help insure that an idea remains in consideration by throwing their whole quota to that idea, but then, of course, they cannot vote for other ideas. The ideas receiving the most votes are transcribed to a new sheet, but no idea having less than a threshold, say six votes, is transcribed (Figure 10-5).

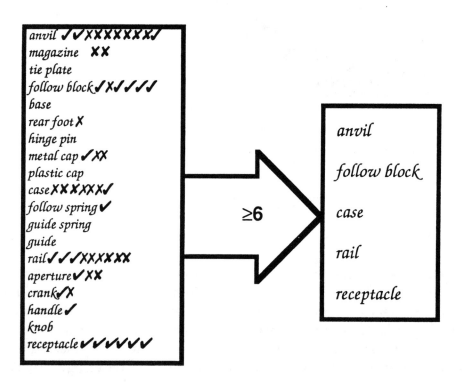

Figure 10-5. Threshold voting is a technique to reduce the number of ideas. Here is a threshold of six, used to reduce the number of machine parts considered for human factors improvement.

If necessary, the group may want to further reduce the new list with another threshold vote. The group can adjust the threshold and quota on each round to ensure that the list grows smaller. Eventually, the group tries to combine the best parts of the ideas on the new list into one idea. This approach ensures wide participation, so nobody feels left out.

10.3.2 *Voting with campaign speeches*

Another way is to vote, with perhaps three votes per person, but keep any idea that gets even one vote. Retranscribe the remaining ideas and have someone who *didn't* vote for an idea elaborate on that idea for one minute. By speaking in support of an idea you didn't vote for, you learn more about the idea, and remove some of the politics from the reduction process. After the speeches, the group does a further reduction, then continues the process by this or another method.

10.3.3 *Blending ideas*

Another good reduction design, perhaps used after one of the above tools has already been applied, is to try blending as many of the ideas as possible into one unified idea that all participants can feel they own. One danger, however, is getting bogged down trying to fit every last idea into one grand synthesis.

10.3.4 *Applying criteria*

If the group has a list of criteria prepared in advance of the brainstorming, the brainstormed list can be reduced by applying the criteria to each idea and eliminating those ideas that don't meet all the criteria. Often in requirements exploration, the criteria list is set by a prior brainstorming and reduction session. And, of course, the final requirements document will become a criteria list against which various design ideas will be tested.

10.3.5 *Scoring or ranking systems*

A more refined method of applying criteria is to develop a weighting formula. This formula gives points for each idea on each criterion, then the points are weighted by the importance of each criterion and summed. The ideas with the largest sums are retained, while the others are discarded. This type of scoring system usually demands a significant amount of work, and is thus usually applicable only when the ideas have been fairly well refined.

10.4 Helpful Hints and Variations

- Brainstorming is not the only idea-generation method. One variant is *brainwriting,* in which each participant starts with a large sheet of paper and writes about an idea for a specified amount of time—perhaps one to three minutes. Then the papers are shuffled and redistributed. Each person reads the idea on the new paper, then continues writing ideas stimulated by that idea. Again, after a time limit, the papers are redistributed and the process repeated.

After about five rounds, the papers are taped to the wall and everyone walks around reading them. Colored markers may be used to "vote" on appealing ideas.

With the proper computer networking, brainwriting can be done on-line, even with the participants not in the same room or even the same country. An on-line brainwriting session can be done in real time, or can span several days, or weeks.

- Another idea-generating approach is to use electronic mail to run a "conference." A topic is established, and the conferees are signed up to have easy access to all the conference material. Any member of the conference can add anything at any time, and it is immediately available to all. Electronic conferencing is a kind of super brainwriting, and may easily produce overload. One preventive is to keep the topics narrow enough, in much the same way ordinary meetings are kept to a single topic.

10.5 Summary

Why?
Ideas are essential, so the ability to gather a group of people on short notice and create a fresh batch is critical to successful requirements work.

When?
Idea-generation meetings are often components of other meetings, as we shall see. A specific idea-generation meeting can be called at any time that two people agree they need some ideas.

How?
Keep in mind the following:

1.   Do not allow criticism or debate.

2.   Let your imagination soar.

3.   Shoot for quantity.

4.   Mutate and combine ideas.

5.   Reduce the number of ideas to manageable proportions after the generation phase, using any one or more of several methods.

6.   Design your own variation on brainstorming, as long as it meets the four basic principles.

Who?
There can't be too many people in idea-generation meetings. As long as they are willing to play by the rules, extra people mean extra ideas. In fact, recruit a stranger for every idea-generation meeting, just to keep the ideas from growing stale.

11   RIGHT-BRAIN METHODS

In order to keep track of where we are in the requirements process, we need maps. Because most effective maps are visual, we'll have twice as many ways of reducing ambiguity if we can mobilize the right side of our brain and thus use visual tools to supplement our left-brain, or verbal, tools.

> Reader: This chapter discusses some right-brain tools. Since "discussing" is a left-brain activity, reading about them isn't exactly the best way to learn this material. Instead, get a set of colored pencils, markers, or crayons, and a large supply of blank paper. Then, after reading about each technique, put down the book, turn off your left brain, and let the right side give it a try. In other words, don't be a reader, but a drawer.

11.1 Mapping Tools

Most mapping tools allow us to produce maps at different levels of detail. Through exploring, we get more and more refined maps, though we don't go as far as Lewis Carroll's German cartographers who eventually made a map so precise that the only way to view it was to lay it out over the countryside it mapped. With the two tools discussed below, we eventually get a map that is close enough to the territory to represent it for practical purposes, which is why our choice of notation must be led by our purposes.

11.1.1 Sketching

Architects and engineers start their design process with *sketches.* Later, they refine the sketches with more and more precise notations. Any development work that produces a physical product is likely to rely heavily on a series of sketches as maps to keep track of progress (see Figure 11-1).

In information systems work, however, developers have not been taught the art of sketching, and too often their work suffers from overly precise notations employed too early in the process. The architect Frank Lloyd Wright used to complain that his buildings were ruined by the "inferior desecrators" who furnished them—that is, those who filled in the details. In the same way, a good sketching notation may be spoiled by attempts to make it more precise.

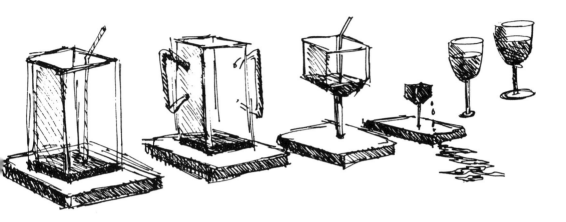

Figure 11-1. The visual evolution of an idea, from raw function to elegant refinement.

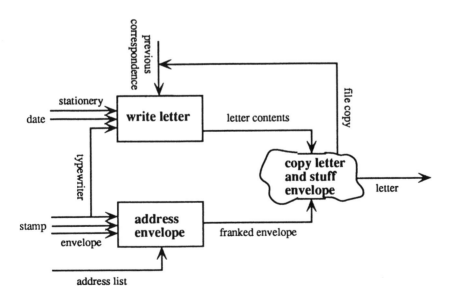

Figure 11-2. A Wiggle Chart of Figure 1-4, showing ambiguity about how the copy will be made and perhaps how the envelope will be stuffed.

11.1.2 Sketching Wiggle Charts

Wiggling is a sketching technique that can be used with almost any visual diagramming notation, with the wiggly lines indicating incomplete precision in part of the diagram. Figure 11-2, for instance, is a Wiggle Chart of the diagram in Figure 1-4.* In this case, the wiggles show uncertainty about exactly how the letter will be copied. Comparing Figures 1-4 and 11-2, we can instantly see the different degree of precision that each implies. This "precision of imprecision" allows developers to focus on just those areas of the map that need further definition.

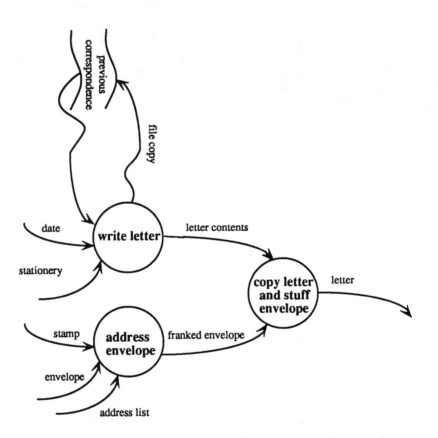

Figure 11-3. A Wiggle Chart of Figure 1-6, showing ambiguity about how previous correspondence will be stored and retrieved.

---

*For a complete discussion of Wiggle Charts, see Gerald M. Weinberg's *Rethinking Systems Analysis & Design* (New York: Dorset House Publishing, 1988), pp. 157-61.

Figure 11-3 is another Wiggle Chart based on Figure 1-6, illustrating that wiggling can be applied to almost any diagramming notation. In this case, the wiggly lines show an ambiguity about how previous correspondence will be stored and retrieved. There might be paper files, electronic media, microfiche, optical disk, or some combined approach. With the wiggles, we need not imply that we know more than we do. Nor do we need to make premature design decisions during the requirements process, just to make our territory fit our map.

11.2 Braindrawing

A visual variant of brainstorming is *braindrawing*. Braindrawing proceeds in much the same way as brainwriting, except that the participants draw pictures and avoid the use of words. Each participant starts a drawing, then passes it to another participant after a few minutes. By adding to each other's drawings, the participants stimulate, and are stimulated to, creative ideas.

After a few cycles, the braindrawings are mounted in an "art gallery," with each drawing presented by someone who didn't contribute to it. Using pictures in this way brings other parts of the brain into the creative process, so braindrawing can often be used effectively in conjunction with word-oriented idea-generation methods.

11.3 Right-braining

In the same way that braindrawing is a variant of brainwriting, we can transform almost any left-brain technique into a right-brain version. Figure 11-4 shows one example of how this can be done. Our objective was to create a new concept in the design of an iced tea glass. Anyone who drinks iced tea realizes that an iced tea glass must have a number of important attributes, many of which are overlooked by restaurants not paying attention to these details.

We put our graphic artist, Marc Rubin, to work on the problem, using a visual variation of other requirements procedures. The goal of this technique is to encourage partial thoughts and ideas to be conveyed as images, by transforming a verbal procedure to a visual one. The visual form may reveal ambiguities that the verbal form concealed.

We first decided on eight important attributes of an iced tea glass and asked Marc to make a spontaneous sketch of whatever visual image came to him from each one.

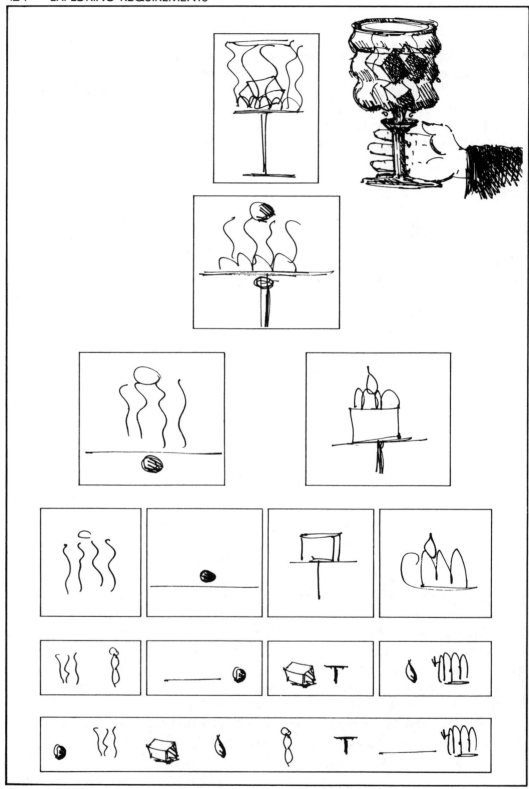

Figure 11-4.   The visual creation of a new concept in iced tea glasses.

The bottom row of Figure 11-4 shows his visual response in turn to the attributes "easy to hold" (a stone), "no condensation" (running streams), "cold inside" (ice cubes), "easy to drink out of" (drop of water), "aesthetically pleasing" (a woman), "very visible tea" (a bold letter T), "easy storage" (a straight line), and "warm hands" (radiator).

After the attributes have been converted into visual images, they are paired at random as shown on the second line up in Figure 11-4. Each of these pairs is then visually condensed into the next image. The only rule is that the condensed image must somehow capture the visual essence of its two constituents. This process is repeated until a single final image is produced, which will probably not look anything like our original ideas of potential solutions. In fact, we may have forgotten the problem that we were originally thinking about through our deliberate effort to force new ways of looking at the problem.

When we reach the top of this visual tree, we tell ourselves that this last image *is* the solution to the problem, the new iced tea glass. We then proceed to refine the image until it actually becomes a solution, taking care to limit our approach to the visual essence of this final image. The solution produced in this session shows an iced tea glass with a "warm" stem designed to be held comfortably in the hand. The radiator worked its way visually into the bottom of the liquid container in the form of an insulated glass bottom to prevent cold hands and condensation on the handle.

Every process we use can similarly be "right-brained," revealing verbal assumptions, stimulating new ideas, and letting our left brain relax for a change.

11.4 Helpful Hints and Variations

- One way to deal with miscommunication arising when two sets of experts participate in the same requirements process is to use more pictures and fewer words. Arguing about pictures doesn't seem to acquire the same violence as arguing about words, and people who don't even speak the same language can often understand and respond to one another's drawings.

- Although sketching is essential to an orderly requirements process, a map can be too sketchy. In *The Hunting of the Snark*, Lewis Carroll's Captain had a map of the surface of the ocean, as shown in Figure 11-5. Of course, since this map accurately described any area of the ocean that didn't contain a land mass, it was widely applicable. It was not, however, useful. To be useful, a map must contain information—enough to make it worth our while to study it, but not so much that our capacity for understanding is overwhelmed. Whenever we draw a map, we should test it against both of these information criteria.

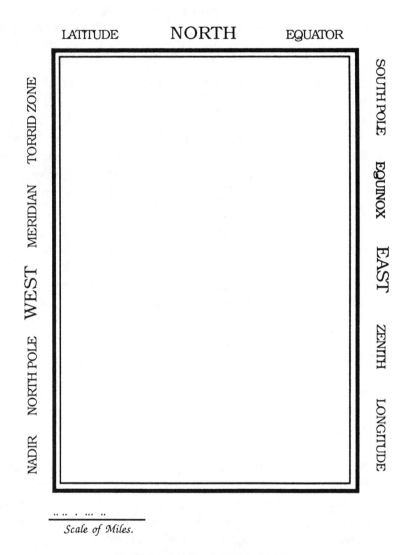

OCEAN-CHART

"He had bought a large map representing the sea,
    Without the least vestige of land:
And the crew were much pleased when they found it to be
    A map they could all understand."

Figure 11-5.   The Captain's map, from *The Hunting of the Snark*.*

11.5 Summary

Why?
"Exploring" comes from the idea that developers work with maps of requirements, not requirements themselves. We explore in order to make maps, and eventually we develop a map that is close enough to the territory to represent it for practical purposes. Visual tools are often the best way to work by successive approximations.

When?
We are always working with maps, not with the territory itself. Therefore, we must *always* be consciously working to ensure common understanding of the various maps we use. The question, "Could you please explain your notation?" is always in order in the requirements process, and nobody should be made to feel foolish or uncooperative for asking it.

How?
Mapping is not so much of a process as it is an awareness and commitment to clear communication. Steps in the process when communication problems arise are the following:

1. Don't leave all the hard parts to be handled by someone else, as in the Guaranteed Cockroach Killer kit.

2. Be prepared for the changes that new methods, especially new notations, will introduce.

3. Accept the differences among people, and don't try to force people to conform, or belittle them for being "stupid." Instead, make sure at every step that everyone can read the map, in order not to exclude anyone.

4. Remember, when you get touchy about notational nuances, remind yourself that the map is not the territory, and that there's no such thing as a perfect translation.

5. Learn to sketch, so that each new map contains sufficient information, not too much information, yet does not imply precision that's not intended.

Who?
Everybody uses maps, though not everybody is equally facile with each system of mapping. Those who are more facile have the responsibility to assist those who are less so, and not in a condescending way.

12   THE PROJECT'S NAME

"Don't stand chattering to yourself like that," Humpty Dumpty said, looking at her for the first time, "but tell me your name and your business."

"My *name* is Alice, but—"

"It's a stupid name enough!" Humpty Dumpty interrupted impatiently. "What does it mean?"

"*Must* a name mean something?" Alice asked doubtfully.

"Of course it must," Humpty Dumpty said with a short laugh: "*my* name means the shape I am—and a good handsome shape it is, too. With a name like yours, you might be any shape, almost."

—Lewis Carroll, *Through the Looking Glass*

Although you may believe, like Alice, that names are arbitrary, Humpty Dumpty had the right idea. The right name increases possibilities, while the wrong name stifles them. Don't you react to the Humpty Dumpty Project differently than you do to the Production Optimization Project?

The name is people's introduction to a project, and is another source of ambiguity. This chapter will present a heuristic for selecting an unambiguous name.

12.1 Working Titles, Nicknames, and Official Names

No matter how a project starts, it almost immediately develops a working title. A good working title unambiguously identifies the project without implying a solution. This name is crucial, for it is the way people are introduced to a project before there is a chance to inform them about the project's true business. Indeed, hearing the name may be the first time people become aware that there *is* a project.

Often, a project develops more than one name, as in our example of the Do Not Disturb Project in Chapter 10. We once worked with the XYZ Company, which was building a software system for the ABC Bank. At XYZ, the project was known as "the Bank's System." At the bank, however, the project was called "the XYZ

System." In this case, the lack of a unique, official name was symptomatic of both parties wanting to disown the project. We forced them to face the ambiguous ownership by running a naming competition among the bank's employees—who were, after all, the client.

Even projects with official names tend to acquire nicknames, like "the Black Hole," or "Don's Project." Although you can learn a lot about how a project is progressing through its nicknames, they can introduce an element of ambiguity. If enough people talk about "Marketing's Project," instead of using the official name, the impact the project might have on other parts of the organization will soon be overlooked.

All project names—from working titles to official names—are dangerous sources of ambiguity, so choose them carefully. When choosing a name for the project, you'll find that making the choice is a cogent way to explore what the project is required to do.

12.2 The Influence of Names

The name of a project can profoundly influence its results as well as the behavior of the participants. Recall in Chapter 10 how the Do Not Disturb Project and other names affected Jack and the others when they tried to generate ideas in their brain-blizzard? Each participant's offering directly related to what they each believed was the project's name and, therefore, its purpose. To illustrate this point, we'll describe a demonstration we've used successfully in our seminars.

12.2.1 A naming demonstration

We divide the audience of professional designers into two groups of about fifty people each. The groups are then asked to organize themselves into teams of four to six people, and told that they will be given a design exercise and half an hour to create a conceptual solution. They are also told that they'll be given an opportunity to present their results to the entire audience. If they do, however, they will be judged by everybody on two factors:

1.   Did they actually produce a solution to the problem?

2.   Is the solution highly innovative?

About twenty teams (ten in each of the two groups) set to work. What they don't know is that the groups have been given slightly different problem statements. One group (let's call it the "Name Group"), gets the first description in Figure 12-1, while the second group (the "Function Group") gets the second description.

Notice that the only difference between the two descriptions is in the way requirement 3 is expressed, by name or by function. When the teams are all finished, they are invited to report if they wish.

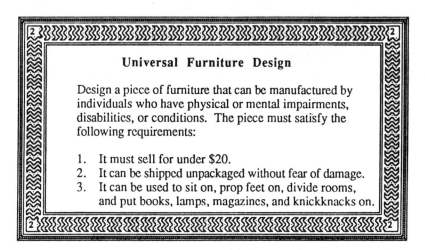

Figure 12-1. Two slightly different problem statements.

Reader: Before reading on, try to predict what happened. In particular, try answering the following three questions.

1. Was there any difference in social behavior between the Name teams and the Function teams?

2. Was there any difference in the number of teams in each group that chose to make presentations of solutions that met their requirements?

3. Was there any difference in the number of highly creative solutions presented, as judged by the audience at large?

Now, check your predictions against the results in Figure 12-2, taken from a typical seminar.

	Name Teams	Function Teams
Social Behavior during the 30 minutes	highly contentious	frivolous
Number of Presentations	3	10
Clever solutions as judged by others	0	3

Figure 12-2. Results of the furniture requirements demonstration.

12.2.2 What naming accomplishes

The furniture requirements exercise shows vividly that how things are named can affect both what is produced and how people behave while producing it. In the demonstration, the two groups worked on different sides of the room, so they could watch each other from a distance without hearing any details. Not only were their behaviors different, but they themselves—while they were working—noticed that there was a striking difference between the two sides. In real life, however, we all may be equally influenced by names but have no group with which to compare ourselves, so we may be totally oblivious to the way we've been influenced by the name.

The project name may be the first name encountered in doing requirements work, but it's by no means the last. We name requirements documents, parts of a product, characteristics of a product, pieces of work, people who take different roles in the project, groups of users, product features, and anything else we can lay our tongues on.

In each case, the name may have been chosen carelessly and arbitrarily, but comes to be used as if it possessed a great deal more precision. Therefore, to create an unambiguous name, we offer the following heuristic. This same tool can also be used to change names, but changing is much harder than getting it right in the first place, because names have a way of sticking.

12.3 The Naming Heuristic

This involves a three-step process: First propose a name, next offer three reasons why the name is *not* adequate, and then propose another name that eliminates these problems. The steps can be repeated until a useful name is developed. We used this process on the Elevator Information Device Project, and it was simple and natural.

Now that we understand more about the influence of names on possibilities, let's delve a bit deeper into why the heuristic works. Suppose we wanted to name the seminar's furniture design exercise. The first possibility that springs to mind is the "Living Room Suite Project." Here is a list of arguments against this name:

1. We may want to use this furniture in other rooms.

2. "Suite" implies that there is more than one piece of furniture, and that the multiple pieces are coordinated in some way.

3. Nothing in this name implies anything about the method of manufacture though the problem statement clearly states that re-quirement.

After considering these shortcomings, we offer a second choice, "Disabled Chair/Table Project," which has the following difficulties:

1. This name strongly implies two particular pieces of furniture, em-phasizing those functions over the others, though nothing in the problem statement implied any more importance to these functions.

2. The word "disabled" may be interpreted to imply something about the people who will use the furniture.

3. Nothing in the name implies the severe cost constraint of selling for under $20.

Having tried these two names and found them wanting, we might turn to a third name, "Cheap Universal Furniture Function Project." We could carry the critical naming process further, but actually never find a perfect name. At some point, though, we'll stop and use what we have because the important thing is *not the name, but the naming*. The point of the naming heuristic is that having gone through this process,

1. we'll probably have a better name

2. we'll certainly have a better understanding of what our real prob-lem is

3.   we'll certainly do a better job of communicating that understanding to other people who join the project later

Even when thinking goes on largely in terms of pictures, the name is what gets transmitted first from one person to another. The people who receive the name will each reconstruct a picture of the name, and none of the pictures will precisely match what was in the first person's mind. (See Figure 12-3.) That's why we say:

One word is worth a thousand pictures.

Figure 12-3. **A picture is transmitted through a translation into a name, which is then translated into another picture, once for each receiver. We say, "One word is worth a thousand pictures."**

12.4 Helpful Hints and Variations

- Although it can be a solitary exercise, naming things is definitely more useful as a face-to-face activity. In addition to achieving the given purpose, a face-to-face naming session has many side benefits. Early in the project, it offers a relatively innocent task as an excuse for the major players to get acquainted. If the players don't act as if it's an innocent task, then the session offers a preview of some of the political issues that may be smoldering under the surface.

- When meeting to explore names, participants need a thesaurus and dictionary handy. The thesaurus helps generate new naming ideas, while the dictionary helps suggest subtle implications of the words that might easily be overlooked. We like to use our personal computer with a word processor, thesaurus, and dictionary on-line.

- Although a naming contest is an excellent idea-generating tool, plan for what you will do if you don't get enough good suggestions. You'll generate a lot of bad feeling if you don't choose a winner, or if you announce, "There were many excellent suggestions, but in the end our beloved president gave us the winning entry."

- One way to proceed is to choose a name and a subtitle. The name can be catchy, possibly an acronym or a word that suggests a nice pictorial symbol. It can even be a pictorial symbol. The subtitle, on the other hand, is the actual working title. For instance, our project might be called CUFF (for Cheap Universal Furniture Function: A Project to Hire the Handicapped).

- Beware of too clever an acronym. In particular, beware of "backronyms," acronyms that were actually chosen for their cleverness before the individual words were chosen. If you start with a clever name and then force-fit words to obtain a working project title, you're almost bound to come up with an inaccurate, ambiguous title that will haunt you for the rest of the project.

- Backronyming, however, can be used as an idea-generating device, if you're stuck. Start with some silly name and then use the thesaurus to hunt for words that might fit both the initials and the actual meaning of the project. If the activity is playful enough, you just might discover something useful.

12.5 Summary

Why?
Names are used repeatedly, and they tend to steer the mind. If they are misleading or ambiguous, they become the starting point for trouble.

When?
Any time a new project or subproject is started, or any time there is anything new to be named, the name must be chosen with care so as not to be ambiguous or misleading.

How?
Follow this cycle:

1. Propose a name.

2. Offer three reasons why the name is not adequate.

3. Propose another name that eliminates these problems.

4. Repeat the naming process until you develop a usable name.

5. Don't go on forever looking for the perfect name. It doesn't exist.

Who?
The naming activity can be done by one person, but is best done face to face in a group consisting of key participants in the project.

13   FACILITATING
IN THE FACE OF CONFLICT

In all the great exploration dramas, there comes a moment when two or more of the principals clash. Should they abandon the canoes and proceed on foot? Should they backtrack through the jungle to rescue Finley from the apes? Should they steal the jewel from the idol's head and risk a perpetual curse?

These are great dramatic moments, but in real-life requirements work, you would be happy to do without them. Unfortunately, all the preparation in the world won't save your project from an occasional intense conflict. It may not be terribly serious, but if it isn't handled right, it could mean a bad end to the project.

Out-of-control conflicts tend to damage projects because they *reduce the tendency to explore possibilities.* Just as in exploring the jungle, most people will run away at the first sign of conflict, rather than stay to see what possibilities are offered. At such times, what you need is another kind of tool: a person who is a skilled *facilitator.* Skilled facilitators can do more than merely cope with conflict. They can turn conflict into a source of new possibilities.

13.1 Handling Inessential Conflicts

The first thing a facilitator does is determine whether a conflict is essential or inessential. An essential conflict concerns *this project, at this time, and involves the people who are present.*

If the conflict is about another project, another time in this project, or people who are not present, then it's not essential. A facilitator resolves an inessential conflict by making everyone aware that it belongs somewhere else. Once the inessential has been stripped away, usually there is no conflict left. If there is, the facilitator can meaningfully apply negotiating techniques, as we'll discuss in Section 13.3.3. The best way to show how a facilitator handles certain inessential conflicts is through some examples.

13.1.1 Wrong time, wrong project

Jerry was facilitating a requirements meeting in which two participants, Morris and Eileen, took violent objection to the mention of a particular programming technique.

"We've tried that, several times," Morris said, with a tone that implied that the rest of us were utterly stupid.

"Yes," Eileen agreed, "we tried it, and it didn't work."

"In fact," said Morris, "it was a total disaster, and killed the project."

"Well," Jerry said, trying to stay calmly in the here and now, "in that case we should know a little more about it. When and where did these disasters take place?"

Morris looked questioningly at Eileen. "As far as I can recall, Eileen, it happened on three or four application projects."

"Yes, I can think of at least four."

"And is this an application project like the others?" Jerry asked.

"No!" Morris said proudly. "This is a systems programming project, which is even harder than an application."

"And why is it harder?"

"Well," said Eileen, "for one thing, we can't use COBOL. This will be programmed in C and Assembler."

"And the other projects were done in COBOL?"

"Of course!" said Morris. "All of our applications are written in COBOL."

"And could there be something about COBOL, for instance, that makes this technique inappropriate?"

"Well, sure! COBOL is entirely different."

Suddenly everyone in the room understood that whatever this conflict was about, it wasn't about a technique that didn't work in COBOL. Once they were brought back to the present, it wasn't difficult to find out what the real conflict was, and to resolve it.

13.1.2 Personality clashes

Perhaps the most familiar example to illustrate the wrong-time-place-people principle is the *personality clash*—when two people always seem to take opposite sides regardless of the subject matter. If the two people involved have a long history with one another, their clash is usually a case of confusing the past with the present. Something from the past probably hasn't been resolved, and the facilitator has to test this hypothesis. When combatants say things like, "You always...," or "You never...," you can be quite sure that this is a continuation of an old argument.

When the facilitator asks them if this is really a here-and-now issue, most of the time they'll come to their senses and decide to settle their personal problems somewhere else. If you have the impression that it's not that easy to handle such a conflict in the real world, that's because you haven't seen a skilled facilitator do it.

If the combatants have little or no history, then you can safely assume that the clash is a case of mistaken identity—or, rather, *two* cases of mistaken identity. Each person somehow sees the other as "just like" someone else in their life, now or in the past. Sometimes these mistakes can be cleared up by recognizing the nature of the problem and providing moderate face-to-face facilitation. At other times, it may be a matter for the psychoanalyst, in which case the best you can do is get professional help.

A personality clash always intensifies when a facilitator attempts to take sides. On the other hand, if nobody can bring the contestants to the here and now, then one or the other must be removed from the project. Quite often, things get better as soon as a person with authority sits them both down and says, "You will both be removed if you don't work this out by the time we leave this room."

13.1.3 Indispensable people

Of course, the threat to remove someone can't be an idle one, so what can you do if both of the people are indispensable? In that case, the remedy is simple: Eliminate both of them.

Figure 13-1. When you have two indispensable people who can't get along, eliminate both of them.

No project can succeed if it depends totally on a person who is permanently out of emotional control. The idea that the project will fail if one person isn't available merely subjects you to continuing extortion by someone who is emotionally unstable.

The secret to handling this type of conflict is to have the courage of your convictions. The sooner you face facts, the better your chances of success. In our experience, nobody's indispensable early in a project. Delay renders a soluble problem insoluble.

This kind of conflict often reveals flaws in the way the requirements work is being managed, which is one reason you may be reluctant to face it. If you really do feel that someone is indispensable, you might take a look at your maps and models, for example, to see why they haven't been understood by everyone.

13.1.4 Intergroup prejudice

Another example of an out-of-time, out-of-place conflict is the kind of generic conflict that takes place between, say, technical people and marketing people, accountants and auditors, or security people and just about everybody else. There are plenty of essential conflicts between engineering and marketing, but quite often one side or both anticipates conflict with the other. They've experienced such conflict before, and because it was so unpleasant and unproductive, they don't want to experience it again. No matter if these are different people, on a different project, at a later time—they start the conflict before the "bad guys" get a chance to start it first.

The best way to prevent this kind of group prejudice conflict is to provide lots of early opportunities for the parties to get to know one another as human beings before they are forced to deal with one another in their professional roles. Informal gatherings, such as pizza lunches, popcorn breaks, or even a party, can do this job—if you structure them so that people don't just sit with their own cliques. Even better is to provide time at the beginning of the schedule for professional team-building activities. Of course, many of the tools used for specific jobs in the requirements process also serve as vehicles for team-building, but be sure that they don't come too late to soften the conflict.

Team-building activities won't prevent all intergroup prejudice. Without warning, someone will start spouting such rubbish as, "You engineers are all alike! You can't let go of anything until it's perfect." or "If you weren't always thinking of how to make a quick buck, you marketing types might get a high-quality product that would sell on its own."

When the participants know each other as individuals, however, the facilitator can call upon that personal knowledge to bring the participants to the here and now. Then the facilitator asks, "Are you speaking about Bob, or some other engineers you've known in the past?" or "What have you seen or heard from Cally that suggests that she is not interested in a high-quality product?"

13.1.5 Level differences

Similar conflicts arise across management levels within an organization. Level n and level n+1 tend to push in opposite directions, which means that a person in a middle level will push differently depending on who is in the room. Sometimes it seems easiest to avoid such conflicts by keeping people at different levels out of the same meeting, but that tactic is a serious mistake. Level differences are most easily handled when conflicts are faced directly with two or more levels represented in the same room at the same time. Everyone is in the same organization, with the same ultimate goal, and the only time it's easy to forget that is when the other people are not visible.

13.2 The Art of Being Fully Present

Did you ever smack your thumb with a hammer, or drill into a water pipe concealed behind a wall? Tools are like that. If you use them without giving them your full attention, you can wreck the work and injure yourself. With interpersonal facilitation tools, you may not injure yourself physically, but you can leave psychological wounds from which a project may never fully recover.

To be a good facilitator, you must develop *the art of being fully present* with the people whose interaction you are facilitating. Of course, nobody can be fully present at all times—distractions are a part of life. However, it's the facilitator's job to notice when it isn't a fully present meeting, and to act. To do this kind of facilitation, you must

1.   Be free to notice everything, and don't pretend certain things aren't happening.

2.   Be free to ask about anything puzzling.

3.   Be free to comment on anything, especially on your own reactions.

4.   Be free to comment when you don't feel any of the above freedoms.

With these freedoms, you can say such things as, "I notice that John is working crossword puzzles, and I find that I'm distracted from the content of the meeting because I don't know what's going on with John. John, can you let me know what's happening for you?" This informs the group that you aren't fully present and also calls their attention to the problem of John not being fully present.

Here are some other types of comments that the fully present facilitator can make:

*   "Although I'm supposed to be neutral at this time, I find myself favoring proposal A because I have no written documentation of proposal B."

*   "I'd like to take a five-minute break to get my thoughts together."

*   "I'm feeling very angry right now, and I'd like a two-hour break to cool off."

*   "I'm unable to follow the meeting when more than one person talks at a time."

*   "Phoebe, I'd like to hear everything you have to say, but I'm not sure if you were finished before Marilyn started talking. Were you?"

Figure 13-2.   Q. What questions does a six-hundred-pound facilitator ask? A. Anything he wants! To be a good facilitator, you must develop the art of being fully present—free to notice, free to ask, and free to comment—but a bit more gently than a gorilla.

13.3 Handling Essential Conflicts

The fully present facilitator models the kind of behavior that's desirable for everyone if your meetings are to be effective tools. People who are fully present help others avoid or derail inessential conflict. But how can the fully present facilitator deal with the essential conflicts?

13.3.1 Reframing personality differences

A common type of essential conflict would be labeled "personality clashes" if the term hadn't already been adopted for the inessential conflict discussed earlier. Such conflicts do not involve just two people, but two groups of people, usually with everyone in the room taking one side or the other. These conflicts don't arise because one professional group stereotypes another. They arise because people even in the same professional group have different personalities, so what meets the needs of one may threaten another.

For instance, some people are more naturally optimistic than others. Some people prefer to wrap up all the loose ends, while others like to leave a few possibilities open. Some primarily want a reward, while others have a greater need to avoid punishment. Some need to feel in control of the situation; others want to be under the control of the situation.

To handle such essential personality conflicts, you as the facilitator start by acknowledging what is going on and then work toward each group's accepting they all have a perfect right to their own *desires,* though not necessarily to getting those desires *satisfied* (Figure 13-3). This acceptance means, for one thing, that everyone must forego name-calling, which most people will agree to when it's made explicit.

Figure 13-3. Meetings run better if everyone understands that we all have a perfect right to our own desires—though not necessarily to getting those desires satisfied.

The acknowledgment and acceptance will be advanced if you can put a positive interpretation on both sides—something that the other side can recognize at least as well-intentioned and reasonable, if not correct. We call this step *reframing.* A facilitator might reframe the conflict as follows: "I hear one group saying they'd like to save time by making this the final document. I hear the other group saying that although they know of no exceptions at the moment, they would like to leave the document open to changes, just in case. Is that a fair statement?"

When you find a reframing of the situation with which both groups can agree, you will definitely have eliminated the name-calling and forced each side to acknowledge the right of the members of the other side to have their own feelings. You'll then be in a position to seek *the third way*—a reframing that gives members of each side at least some of the safety they need. In our example, both sides agreed that the document could be made final, and they adopted an amendment procedure designed to be used in case something unforeseen came up. However, the procedure was not so easy to use that it would be abused with frivolous amendments.

13.3.2 Negotiating

Obviously, the effective facilitator must understand negotiations, and be able to lead negotiations in which the participants may not be very skilled. In a meeting between a service bureau and some of their customers, the "optimists" (customers) argued that there could be no errors in the input data, so they wouldn't pay for error-handling programs. The "pessimists" (service bureau people) simply could not accept this assumption.

The two sides argued to no avail until the facilitator observed, "You are arguing about a *fact*, a fact that cannot be known until the future. Each of you has experiences in the past to make you believe that your prediction about input errors is correct, but nobody will really know until after this system is built." This statement brought the argument to a halt, with both sides turning on the facilitator.

Now, if you can't stand people turning on you when you make true observations, you won't be very good at facilitating requirements meetings. Every requirements process we've ever facilitated has had at least one argument about facts in the future. In this case, the facilitator stilled the attacks by proposing, "Since we don't know how many errors there will be, suppose we set a price to be paid to the service bureau for each error in the input. For example, since these customers believe there won't be any errors, we can set the price at the cost of running the job again, after the error has been corrected."

In the face of large potential rerun costs, the customers decided that perhaps they didn't really know that there couldn't be any errors. They accepted the service bureau's offer to assist in conducting an error study, and eventually agreed to pay for programs to handle the five most common errors automatically. They agreed that any other errors would result in a paid rerun, or acceptance of the faulty output.

What the facilitator did by this proposal was to begin a process of *revealing and testing assumptions*. Sometimes people get angry when you do this, but less angry than when they find out later that their assumptions were wrong. Over the years, we have learned to welcome this kind of conflict because it is a powerful tool for revealing contradictory assumptions. Of course, if we didn't know how to handle it, we wouldn't welcome it.

13.3.3 Handling political conflicts

Once you've stripped away the inessential conflicts, the true personality differences, and the differences in information, you're left with the essential political conflicts that exist in all development projects. Some of the classic conflicts are

- quality versus planned development time
- cost versus planned development time
- cost of quality versus planned development time
- project payoff versus actual development time
- security versus access
- full functionality versus reliability

Don't be discouraged if you run into one of these essential conflicts in your requirements work. In fact, if you don't reach these essential conflicts, you don't yet understand your problem. During the requirements phase, the facilitator's job in the face of such essential conflicts is easy. In fact, it consists of only two parts:

1. Keep everybody calm by reminding them that these are perfectly natural conflicts.

2. Convince everybody to resolve these conflicts in the design phase.

The resolution of such essential political conflicts is the job of design, not requirements. The requirements job is to state, for example, what level of quality is required and what development time is allowed. When the design work is done, there may not even be a conflict between these two requirements, in which case everyone would have wasted precious breath arguing about them during the requirements process. Not only that, but if someone *wins* the argument, you may wind up compromising quality unnecessarily because the true desires were omitted from the final requirements document. Any compromise made in the requirements stage is no longer available for the designer to attempt to resolve in a creative way.

13.4 Helpful Hints and Variations

- There are times when it's worthwhile to retain a professional facilitator, who can offer several benefits. Experienced facilitators have lots of practice to polish their facilitation skills, and they have no personal stake in one outcome or another—so it's easier for them to avoid getting embroiled in any conflict. They can serve as trainers for your own facilitators, who may be talented but not very experienced.

- Although we are both professional facilitators, we rarely take on facilitation work where we don't have the opportunity to train one or more of the client's people in facilitation skills. That's because local facilitation has many advantages over retaining outsiders. First and foremost is that local people can be available throughout the life of a project, even when meetings are called suddenly. Indeed, it's those sudden meetings that are most in need of facilitation because they're likely to lack planning and yet be held in an emotionally charged situation. So even if you hire outsiders, consider using them to train your own facilitator corps.

  We encourage our clients to "stockpile" good facilitators. They start by identifying people who have already displayed some native talent for this kind of work, regardless of their technical background. Technical knowledge is not essential for facilitators, and in many cases, it's an advantage to the facilitator not to know too much about the technical content of meetings. When they do, they get more easily embroiled on one side or the other of a conflict, even when the technical content is merely a mask for some inessential conflict.

  We then create a volunteer facilitator corps, or support team, distributed as widely throughout the organization as possible. These volunteers take outside training, observe and critique one another, share experiences, and volunteer their services. In a short time, the demand for their time greatly exceeds their availability, because people rarely understand just how useful a skilled facilitator can be until they have experienced one. After they have experienced one, they won't waste their time and emotional energy trying to get along without one.

- Good facilitation is like good police work. When it's done well, you may not notice that anything is being done at all. That's why projects sometimes fail to appreciate the people who are facilitating their meetings. If everyone has some training in facilitation skills, the facilitators' jobs will be easier, because they will have a more cooperative and appreciative audience.

13.5 Summary

Why?
Every project will experience conflict, and it cannot succeed consistently without some kind of facilitation. Some projects are lucky enough to have "amateur" facilitators who can act when the need arises. But if you don't want to depend on luck, develop a corps of trained facilitators.

When?
It's never a good idea to conduct any requirements meeting without a facilitator, because you can't predict when intense emotions will erupt. A good facilitator, however, will prevent most of the inessential conflict, and minimize the rest.

How?

This chapter is intended to stress the importance of facilitation, not to make you a skilled facilitator. There are many good books about facilitation.*

Who?

Every participant should take responsibility for facilitating meetings, but a few people should be professionally trained as specialists in the job, which has its own highly developed skills.

---

*Here are two of our favorites: Virginia Satir, *Making Contact* (Berkeley, Calif.: Celestial Arts, 1976); and David Kiersey and Marilyn Bates, *Please Understand Me: Character & Temperament Types*, 4th ed. (Del Mar, Calif.: Prometheus Nemesis Book Co., 1984).

PART IV CLARIFYING EXPECTATIONS

Some time has passed since the first meeting on the Superchalk Project. Barbara, Larry, and Todd, the design team from BLT Design, are back in the chalk cave with Byron, Wilma, and John. The BLT designers now know that Byron, President of Cheap Chalk Corporation (CCC), is their customer, and that John and Wilma are representatives of two different user populations. They understand how much Byron is willing to spend, and that this project has been initiated because CCC has discovered a new vein of super-pure white chalk, which they wish to market in a new, distinctive form.

Larry: As you know, Byron, BLT Design is the world leader in the design of writing instruments.

Byron: That's why we hired you to design Superchalk.

Larry: Of course. At this stage in the requirements process, we need to find out exactly what you expect Superchalk to be like. Now, over our many years of designing writing instruments, BLT has developed a standard list of attributes that every writing instrument should have.

(Larry unrolls a sheet of flipchart paper, tapes it to the chalkboard, and takes out a felt-tip marker pen. Byron winces but doesn't say anything.)

Larry: Here's our list. A writing instrument should be user-friendly, profitable, manufacturable, easy to sell, innovative, unique, strong, reliable, safe, nontoxic, nonallergenic, clean, easy to package, versatile, appropriately erasable, and should produce easy-to-read writing. Don't you agree?

Byron: Hmmm.

Barbara: Good. And Wilma, what do you think?

Wilma: Sounds wonderful.

John: I can't see anything to add.

Barbara: Good, then we all agree on the attributes.

Todd: I can't wait to get started on the design!

14 FUNCTIONS

The BLT team may be ready to start designing a product, but they're going to be in a heap of trouble if they don't have a clearer idea of precisely what CCC wants from Superchalk. They can proceed to get this information in an orderly way by moving from functions, to attributes, to constraints, to preferences, to expectations, as we'll do on several projects in the next five chapters.

14.1 Defining a Function

Frequently, the first step in this orderly process of gaining information is to define *functions*. Functions are the "what" of a product, describing what the product is to accomplish. They are verbs, representing actions for which the product is the subject.

14.1.1 Existence function

The first function of any system is *to exist*, which at this early stage is not a certainty, but an assumption.* As we know, this assumption is at the root of the decision tree. For example, the client says:

- We want Superchalk to exist.
- We want the Elevator Information Device to exist.

The development of the problem statement starts with this existence function and proceeds by defining what the product must do to exist for the client. For example, "Superchalk will exist when it writes on slate and satisfies our customers." "Writes on slate" and "satisfies our customers" are both desired functions of Superchalk, deriving from the function "exists."

---

*Many of the deep principles of systems design can indeed be derived from the necessity of survival. See Gerald M. Weinberg and Daniela Weinberg, *General Principles of Systems Design* (New York: Dorset House Publishing, 1988).

14.1.2 Testing for a function

To test whether a requirement is actually a function, put the phrase "We want the product to. . ." in front of it. Alternatively, you can say, "The product should. . ." For example, rephrase the previous statements as,

- Superchalk should write on slate.
- Superchalk should satisfy our customers.

To take another example, "display current floor" is a function because we can sensibly say:

"The Elevator Information Device should display the current floor."

"Purple" is *not* a function, because we cannot sensibly say,

"The Elevator Information Device should purple."

To make sense, the sentence would have to read,

"The Elevator Information Device should *be* purple."

or

"The floor display of the Elevator Information Device should *be* purple."

Thus, "purple" is not a function; it is an *attribute* of the system or of some function of the system, as we shall see in Chapter 15.

14.2 Capturing All and Only Functions

Since there are many books on requirements that do a good job of teaching us how to describe functions clearly and accurately, our task here is to concentrate on methods that

1. capture all the functions that the client wants
2. understand those functions
3. don't capture functions that the client doesn't want

14.2.1 Capturing all potential functions

The first step in the process is to have the client brainstorm all conceivable functions to be performed by the system. The designer should not contribute ideas at this stage, but simply facilitate the client's imagination in dreaming up wishes. Several questions may be used to prime the brainstorming process:

- How would you (the client) like to use this product?
- What is the purpose of this product?
- How would others like to use this product? (Refer the client to the user list.)
- Don't worry about how much it would cost. What would you like?
- Don't worry if it can really be done. What would you like?

Using such a brainstorm, the client dreamed up the following functions for the Elevator Information Device:

1. display current floor
2. display the selected floor stops ahead
3. display relevant outside weather conditions
4. provide local 24-hour weather forecast
5. give the floor selection of passengers
6. perform a security check for secured floors
7. de-select for inadvertent selections
8. furnish a priority override for emergencies
9. give appropriate passenger information in emergency situations
10. display directory information
11. give directions for moving and delivery people
12. supply current information to emergency personnel
13. "scream" when passengers are being assaulted
14. include measures to render escape impossible for assaulters
15. give a clear warning that escape is impossible to potential assaulters
16. display relevant information to building occupants, such as
    - individualized tickler files
    - stock and bond quotations
    - sport scores
    - hot news happenings
    - local communications to and from the landlord
    - billboard announcements to and from any occupant
    - current time and date
    - historical events of the current date

- health reminders
- warnings when a passenger is in a detectable state of poor health

17.   provide passengers with the ability to control doors when necessary

18.   have the ability to reserve a specified elevator for such use as moving

19.   include a tilt control for elevators with articulated floors (floors that slant up and down) so that heavy objects may be rolled on and off easily

20.   provide special information for elevator inspection, such as
- date of elevator's last inspection
- date of law violations for each elevator
- other specifics of each violation

21.   give expected number of stops and time of arrival for long runs

14.2.2 Understanding evident, hidden, and frill functions

Once the functions have been listed, they can be organized into three broad classes:

E   for *Evident*
H   for *Hidden*
F   for *Frill*

Evident functions are those to be performed in a manner that is as visible, or evident, to the users as possible, and *hidden* functions are to be as imperceptible to the user as possible. *Frill* functions are those that the client would like, but not if they cost anything, either directly or in compromises with other functions.

To illustrate the differences between **E**, **F**, and **H** functions, consider the design of a teakettle. In traditional teakettle design, the problem is to heat water to the boiling temperature. Since we don't really care *how* the teakettle boils the water, so long as it is quick and efficient, boiling the water is a *hidden* function. But we do want to know *when* the water reaches the boiling point, so informing us is an *evident* function.

A third important function in teakettle design is aesthetics. We may not buy a teakettle that won't look good on our range or shelf, in which case beauty when not in use is another evident function. Thus, for these three functions, the traditional teakettle (Figure 14-1) may be described as

Teakettle #1:
Boiling water (**H**)
Indicating when boiling temperature is reached (**E**)
Looking beautiful to potential buyers (**E**)

Figure 14-1. The traditional teakettle (#1) is described by the assignment of H, F, and E functions.

But this assignment of function types is merely the traditional one. Suppose we made a different assignment:

Teakettle #2:
Boiling water (**H**)
Indicating when boiling temperature is reached (**E**)
Looking beautiful to potential buyers (**F**)

Figure 14-2. The commercial teakettle (#2) is described by a different assignment of H, F, and E functions.

Teakettle #2 has no requirement for attractiveness, and might be a commercial teakettle, for institutional cooking (Figure 14-2). In fact, teakettle #2 might not even exist at all in its water-heating form when not in use. It could be a hiker's teakettle that rolls into a tiny ball to be carried in a backpack.

A third assignment might be

Teakettle #3:
Boiling water (**E**)
Indicating when boiling temperature is reached (**E**)
Looking beautiful to potential buyers (**F**)

Figure 14-3. The Rube Goldberg kinetic art teakettle (#3) is described by still a different assignment of H, F, and E functions.

This assignment produces rather different design ideas for a teakettle. For instance, we might produce a Rube Goldberg kinetic art teakettle (Figure 14-3)—an elaborate glass device that shows off the water in a fascinating manner while it is being heated. The device is designed to scream, "Look everybody, I'm heating water! Isn't it fantastic?" On the other hand, when not in use it fades into its surroundings.

A fourth assignment might be

Teakettle #4:
Boiling water (**H**)
Indicating when boiling temperature is reached (**H**)
Looking beautiful to potential buyers (**H**)

Figure 14-4. Another assignment of H, F, and E functions produces something we don't even recognize as a teakettle. It is sometimes referred to as the "office teakettle." (#4)

Teakettle #4 might be a teakettle for office use (Figure 14-4). It has no need to inform us when boiling because we want the water always to be boiling, or boiling within an instant of when we request it. It need not be attractive, and in fact we would prefer it not to disturb our work in any way, when boiling water or not. Teakettle #4 might also be a microwave oven.

14.2.3 Identifying overlooked functions

Classifying the functions list into hidden and evident functions helps to identify overlooked possibilities because it focuses on how some of the most essential functions of a system are taken for granted. For instance, few people were consciously aware of the information function of an elevator until automatic elevators actually began replacing operator elevators.

Our own example of the Elevator Information Device attempts to restore many of the information functions that were provided in a hidden fashion by the human operator. Just look at some of the functions on our list:

- Display directory information: You could always ask the operator where anything or anybody was in the building. Moreover, the human operator would update information dynamically, so you could find out if Mr. Quackenbush was not in the building at all, or visiting someone on the fourteenth floor.

- Display special information for elevator inspection: The operator was always the first source for an elevator inspector to ask about any unusual events.

- Scream when passengers are being assaulted: Come to think of it, there didn't seem to be any muggings on human-operated elevators, but nobody noticed that the "protect from muggers" function was missing until the operators were replaced.

- Display relevant information to building occupants: Those of us old enough to recall working or living in a building with an elevator operator truly miss that storehouse of information. Not only could Ralph, the operator, supply the current weather and weather forecast, but also the latest sport scores and significant stock market events. He provided a lending library for newspapers (people gave him theirs when they arrived at work at the end of their train ride). If you didn't have time to read, Ralph would recite all the headlines between floors. When the landlord had some message for the tenants, Ralph took care of it. If you wanted something to reach the landlord, you could always count on Ralph to deliver it.

When the hidden functions are named, as in our example, the chances increase that clients will consider those functions they now take for granted, and the chances decrease that when a product is delivered they will say, "How come you didn't. . .?" or "But I assumed it would. . ."

After classifying the functions, you may want to do a second brainstorm to augment the list, this time focusing on functions that were taken for granted the first time around.

14.2.4 Avoiding implied solutions

Another sort of overlooked possibility arises from accidentally implied solutions disguised as function descriptions. When we listed "display current floor," we might not have meant to imply a visual presentation, but most people will interpret this function that way. The function "display current floor" might have been better expressed as "provide current floor location," freeing the designer's thinking from the usual display function. But if we haven't been clever enough to say "provide," rather than "display," we may still be rescued from our assumptions when we ask, "What would it mean for 'display current floor' to be a hidden function?"

The classification question makes us consider the possibility of conveying information in the Elevator Information Device on an unconscious or subliminal level rather than the traditional conscious, and usually visual, level. A variety of innovative ideas immediately spring to mind, such as

- slight vibration patterns in the elevator's tracks

- airflow differences
- odors ("Oh, this is the popcorn floor!")
- colors
- tones
- musical chords
- combinations of several signals

By explicitly stating and classifying function, you're resolving ambiguities and exposing hidden possibilities. Continue to revise the function list as this kind of information emerges. Convey the idea that the function list is like modeling clay, out of which we will sculpt the final function list.

Is it really important to go to so much trouble to avoid implied solutions? After all, won't all ambiguities eventually be resolved, one way or another? Of course they will. You'll resolve them implicitly as part of your design process, and the resolutions will be based on your own experiences, biases, and preconceived ideas, with little or no client participation. The end result will be more costly, less responsive, and less imaginative. As the commercial says,

"Pay me now or pay me (a lot more) later."

14.2.5 The "Get It If You Can" list

When we propose to designers that they brainstorm functions with their clients, we often meet strident resistance. These experienced designers understand that each additional required function increases the difficulty of the design problem, so they certainly don't want to do anything that will encourage the client to dream up more functions that they expect but aren't willing to pay for.

On the other hand, experienced designers also know that if you try to suppress the client's expression of wishes, the suppressed wishes always come back to haunt you later. It may not be easy, but it's far better to get all the wishes out in the open as early as possible, and begin to introduce a sense of reality.

Once all the brainstormed functions are listed, ask the clients to select those that are frills. These are transcribed to another list, which will go into the final requirements document under the title of "Get It If You Can." Heading the "Get It If You Can" list is the following statement:

The designers will be alert to opportunities to provide these functions whenever they can do so without trading away other functions or attributes to get them.

In a rather amazing way, putting the frill functions on the wish list defuses arguments over keeping or discarding someone's favorite idea.

14.3 Helpful Hints and Variations

- As a prologue to the brainstorm for functions, make sure that everybody understands what a function is. Once you start the brainstorm, however, don't stop to correct someone who offers something other than a function, such as an attribute. Just write it down. Later, you can remove it from the function list and use it to start the brainstorming for attributes.

- Figure 14-5 shows a set of ordinary crescent wrenches. Notice how the length of the handle is both an *evident* way of selecting the right size, and a *hidden* way of protecting against too much torque. The angle of the handle, however, is a kind of *translucent* function: People can't articulate why it's bent like that, but when they need it they use it properly. Their body knows why it's there, but their mind doesn't get the message. For some products, you may want to add a fourth function category for translucent functions.

Figure 14-5. In this set of crescent wrenches, the handles demonstrate hidden and evident functions in their lengths, but also demonstrate a translucent function in the angle.

- *Opaque* functions exist when certain user constituencies *must not* know of the existence of the function. For example, many security and national defense systems must have opaque functions. An interesting case is systems for automated credit card verification. They must be available 24 hours a day, but sometimes the machines fail or are taken out of service for maintenance. The opaque function is to keep anyone from knowing when the system is down, which is needed to discourage credit card fraud.

This opaque function can be accomplished by a cheap additional system that simulates the real system for periods of down time, and which automatically approves all credit requests. Since nobody can tell at the credit application station whether the real system is functioning, thieves or deadbeats cannot risk presenting their card, but no honest person need be afraid, nor will be turned down. You may want to add "opaque" to the categories for clarifying functions.

• When reading about providing floor information by odor, our colleague Ken de Lavigne lamented, "We won't start getting innovative features like these until we stop buying from the lowest bidders!" This suggests the Chinese proverb,

"The best is the enemy of the good."

Attempting to optimize one aspect of the product—be it cost, speed, or size— turns attention away from innovations that might in the end produce an even "better than optimum" solution for that very same aspect. So beware of "optimization statements" and heed their warning to slow down and smell the popcorn.

14.4 Summary

Why?
Use the function heuristics to identify the true functions desired, reduce chances of overlooking important implicit functions, provide insights and opportunities for new functions, and create a consistent handling of functions.

When?
Apply the function heuristics at each cycle of refinement, that is, each time you decompose a function into more detailed functions. The first time, of course, is when you decompose the "exists" function and define what the product must do in order to exist at all.

How?
Perform these steps with your clients:

1. Through brainstorming, develop an initial list of potential functions.

2. Classify each function as evident, hidden, or frill.

3. Using this classification, try to uncover unmentioned hidden functions, perhaps by brainstorming to augment the function list.

4.   As you make the classification, look for functions with wording that implies some constraint on solutions and transform the wording to become problem statements, rather than solution statements.

5.   Create a "Get It If You Can" list for the frill functions.

15   ATTRIBUTES

Attributes are characteristics desired by the client; think of them as adjectives or adverbs. Two products could have exactly the same functions, but their attributes can make them entirely different products. A Rolls Royce has more or less the same functions as a Ford, but many, many different attributes.

15.1 Attribute Wish List

At an early stage in the design process, lead clients in a brainstorming session to produce a wish list of attributes. Like the function list, the attribute list is not to be constrained at this time by any thought of how an attribute could be satisfied, nor by any possible conflicts among the attributes.

When BLT Design worked with CCC, for example, the designers didn't brainstorm a list with the client, but offered their own list, based on their own experiences as writing instrument designers. Their list included

user-friendly	profitable	manufacturable
easy to sell	innovative	unique
strong	reliable	safe
nontoxic	nonallergenic	clean
easy to package	appropriately erasable	versatile
producing easy-to-read writing		

It's not a bad list, but the BLT designers may be in trouble. They weren't good listeners, and were impatient to get their clients out of the way so they could started designing. That often happens when an experienced, professional design team works with clients who haven't done much design work. Of course, the BLT design team knows a lot about the design of writing instruments, but they don't know everything about the client. They need to learn one of the first principles of product design:

No matter how many clients you've served, the next client is different.

That's why it's better to have the clients themselves produce a wish list in a brainstorming session. If they omit an attribute that you think is essential, you can always suggest adding it to the list after the brainstorm. Otherwise, they may be intimidated, and bring up their true desires only *after* the project has gone to the devil (Figure 15-1).

Figure 15-1. No matter how many clients you've served, the next client is different.

Here's an example of an attribute list brainstormed for the Elevator Information Device:

flexible presentation	easy-to-select information	polysensory
monosensory	inexpensive	easy to use
having few options	portable	continuous operation
highly reliable	brown	easy to service
providing high-value information	easy to understand	anchorable
soft	compact	lightweight
offering a help facility	nonhazardous	user-modifiable
entertaining	waterproof	buoyant
multilingual	fast response	slow response
consuming little power	having low maintenance	long-lived
self-contained	modular	easy to install
well documented	self-apparent	fireproof
saltwater proof	easy to cool	rugged
reliable	quiet	ergonomic
easy to hear	easy to taste	fun to feel

looks good	sexy	aromatic
having a portable energy source	theft-proof	multipowered
colorful	expensive	safe
legal	sensual	small
large	removable	foolproof
tamperproof	fail-safe	efficient
self-diagnosing	versatile	adaptable
highly innovative	environmentally robust	

15.2 Transforming the Wish List

In our example, we produced a list covering everything anyone might ever want to have as a quality in the elevator system. To accomplish that, we had to keep the brainstorm free of excessive precision. Now that we have the full list, however, we must reduce the brainstormed list of attributes to a reasonable list of what we really want in the system, excluding what is not valuable. But before we can hope to do this, we must understand what the attributes *are* and remove all ambiguity.

15.2.1 Distinguishing between attributes and attribute details

In any list of desired attributes, whether obtained by brainstorming or questioning or from a written requirement, there is generally a mixture of *attributes* and *attribute details*. Attributes are *dimensions* in which we are interested. Thus, "longevity" is an attribute.

Attribute details are points, or sets of points, along that dimension. "Ten years" and "longer than the elevator lasts" are two possible details on the attribute dimension of "longevity." ("Ten years" is actually a single value, whereas "longer than the elevator lasts" represents a range of values, but both are details of the longevity attribute.)

Alternately, "color" could be an attribute and details could include "brown, black, red, or white." If psychologists observe that certain colors tend to discourage vandalism in elevators, there may be a relationship between "longevity" and "color." Thus, different attributes may be interrelated and dependent, though sometimes their relationships will surprise us.

15.2.2 Uncovering attribute ambiguity

Either attributes or attribute details may contain subtle ambiguities, which must be carefully removed. For example, the attribute detail "low maintenance" sounds clear enough at first. But what is the attribute to which this detail belongs? Is it "maintenance cost" or perhaps "time spent on maintenance" or "operational time lost to maintenance"?

Although we often hear of bottom-line thinking and might assume that everyone understands its meaning, the attribute "cost" is perhaps the most ambiguous of all possible attributes. Does it mean "capital cost to build," "operational cost per time period," "operational cost per transaction," "total lifetime cost discounted to today's dollars," "personal cost to user X," "personal lifetime cost to developer Y," "replacement cost," or just what? Instead of making assumptions, we must increase our understanding until all ambiguity is removed from "cost."

15.2.3 Organizing the list

One way to improve our understanding of the attributes is to transform the brainstormed list by the following two processes:

1.   If an item is an attribute detail, write down the attribute(s) it represents.

2.   If an item is an attribute, collect the attribute details that belong to it.

After all attributes and attribute details have been matched, put them into the format:

Attribute = (list of attribute details)

For example, when the BLT designers tried to work out their attribute list with Byron, CCC's president, they reduced the list to three attributes, with their associated details, not all of which were in their original list:

Manufacturable = (cylindrical, pure chalk)

User-friendly = (nontoxic, reliable, nonallergenic, safe, clean, appropriately erasable, versatile, producing easy-to-read writing)

Profitable = (innovative, easy to sell, easy to package)

For Byron, manufacturability involves two attribute details that BLT Design never thought of, *cylindrical* and *pure chalk*. (See Figure 15-2.) This is because CCC's existing machinery can't work with any other material, and can only produce cylinders. BLT Design didn't know anything about chalk manufacturing, but the attribute definition process helped its designers learn—fast.

Here are some examples of attribute definition derived from our brainstormed list for the Elevator Information Device:

Number of senses involved in communication = (polysensory, monosensory)

Cost = (inexpensive)

Response time = (fast, slow)

Power consumption level = (consuming little power)

Ease of use = (flexible presentation, easy-to-select information, ergonomic, easy to hear, easy to taste, well documented, self-apparent, easy to understand, easy to use, having few options, offering a help facility, foolproof)

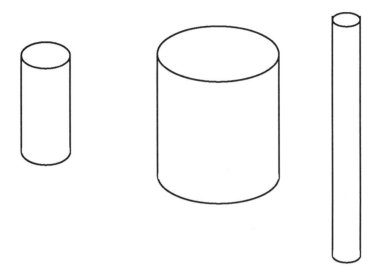

Figure 15-2. To be manufacturable, Superchalk must be cylindrical and of pure chalk. Here are some possible designs of Superchalk, satisfying these attributes.

15.2.4 Discovering insights from the transformed list

Organizing the list like this produces a number of insights. For instance, if there are many attribute details under "ease of use" and very few under "cost," this may mean either that the main emphasis for the system will be on ease of use, or that the client simply doesn't readily think in terms of cost.

With the list in this redefined form, participants almost always notice ambiguities that were not previously evident. For instance, when we try to place "having few options" under "ease of use," someone might argue that systems with "many options" are easier to use. Eventually, we may discover that the more appropriate detail under "ease of use" would be "appropriate number of options."

Also, once the attributes are defined in this way, participants invariably become aware of additional attribute details. For instance, upon seeing

Response time = (fast, slow)

one participant insisted on adding "consistent response," and another added "psychologically appropriate response," giving the more refined attribute,

Response time = (fast, slow, consistent, psychologically appropriate)

15.3 Assigning Attributes to Functions

An attribute does not stand alone, but is rather *a modifier of one or more functions.* An attribute like "manufacturability" for Superchalk might modify the existence function—simply "being." Similarly, an Elevator Information Device might have a desired attribute, "cost," or "safety," that cannot be applied partially. Such cases are attributes of the entire product, and thus of *every* function of the product.

15.3.1 How attributes can modify functions

Other attributes will qualify only certain functions of the system. "Color" may be an attribute of the "display directory information" function, but perhaps not of the "scream when passengers are being assaulted" function.

The same attribute may *qualify more than one function.* "Reliability" is an attribute that could apply to the "display directory information" function, the "display special information for elevator inspection" function, and the "scream when passengers are being assaulted" function.

In short, we need to amend our list of attributes so that each attribute refers to a function, in the form

Attribute A of function F = (list of details)

For instance, we might transform the reliability attribute into the following three components, each associated with a different function of the Elevator Information Device:

Reliability of the display directory information function = (infrequent misinformation, ...)

Reliability of the display special information for elevator inspection function = (tamperproof, high probability of detection, ...)

Reliability of the scream when passengers are being assaulted function = (tamperproof, moderate level of false alarms, continuous operation, fail-safe, ...)

15.3.2 Gaining insights from the new format

Once an attribute is in this form, new insights emerge. For example, when viewing the reliability of the "scream when passengers are being assaulted" function, people recognized several attribute details that had not appeared in the brainstormed attribute list. Adding these details to the list gives a more completely defined attribute:

> **Reliability of the scream when passengers are being assaulted function** = (tamperproof, continuous operation, fail-safe, moderate level of false alarms, very high level of correct reporting, reduces mugging attempts by 80 percent, ...)

15.4 Excluding Attributes

At this point, one attribute may now have become many, thus expanding the attribute list, yet the list may still have many conflicts, overlaps, and ambiguities. During the design process, of course, the designer will resolve all these difficulties, one way or another. To save time and money, however, you need to resolve as many as possible, as early as possible. It is best to resolve them all now, while still trying to define the requirements.

15.4.1 Categorizing into must, want, and ignore attributes

As a first rough cut, the designer and client can attempt to put each attribute in one of three categories:

> **M**   something the client *Must* have
> **W**   something the client *Wants* to have
> **I**   something the designers can effectively *Ignore*

For instance, Byron might assign Superchalk's three attributes to the following categories:

> **(M) manufacturable** = (cylindrical, pure chalk)
>
> **(W) profitable** = (innovative, easy to sell, easy to package)
>
> **(I) user-friendly** = (nontoxic, reliable, nonallergenic, safe, clean, appropriately erasable, versatile, producing easy-to-read writing)

This assignment means that Byron isn't looking for something as fancy as the BLT designers had hoped or imagined. Perhaps they were misled by the project name, "Superchalk," into thinking he wanted something grandiose (Figure 15-3).

Figure 15-3. Explicit listing and classification of attributes helps everyone not to be misled by grandi-ose names.

"Superchalk" was merely a marketing ploy, designed to conceal the fact that this was going to be more or less the same old chalk design. The big change was that CCC would use purer material from their new vein. Byron's only real concern is that CCC be able to manufacture Superchalk using their existing machinery, which can only produce chalk cylinders.

15.4.2 Implicit versus explicit elimination of attributes

Although the BLT designers were disappointed to have a smaller assignment, they had a far easier job than the designers of the Elevator Information Device. With 3 different categories and 300 attributes, this project would have 3^{300} possible assignments, each producing a somewhat different design approach.

Sometimes we work with a client who says that *every* attribute is an **M**. Such a client expects a free lunch and will be disappointed in *whatever* design is produced, unless previously educated in the facts of life by an attribute exclusion exercise.

Out of necessity, then, you must struggle to reduce the number of **M** and **W** attributes, in order to exclude as many attributes as possible that have no substantial value to the client. If you don't, the cost of design will rise astronomically, which is why designers *instinctively* and *implicitly* eliminate hundreds of attributes from consideration.

The attributes that survive this elimination will be the driving force for design. Though designers' instincts are frequently correct, implicit elimination unfortunately also drives out most truly innovative design solutions. Anything eliminated now will have little chance of being resurrected until much too late.

That's why it's essential to generate large numbers of attribute ideas out of the world of fantasy and opportunity—to produce a statement of all attributes in some ideal system without regard to achievability. Only then is it safe to go through our processes of explicit elimination to produce a realistic list of functions and their attributes for the next stage in the process—applying constraints.

15.5 Helpful Hints and Variations

- In the attribute-generation process, encourage *playfulness*. If this process gets too serious, that means people aren't stretching their minds sufficiently.

- If the generation process is not sufficiently playful, remind participants of the elimination step, which gives them the safety to mention "ridiculous" attributes. If people know the truly ridiculous will ultimately be eliminated, it will be easier for them to loosen up.

- In practice, some of the "ridiculous" attributes do indeed open entirely new paths to innovative designs, and so remind participants that even one idea that increases the value of the design by one percent will pay for many "wasted" minutes playing with "ridiculous" ideas.

- Remember that guiding the explicit elimination process is what the client *really* wants, and do not be too greatly influenced by your ideas of what's achievable. In fact, if you're sure it's achievable at this stage, you haven't created enough attributes. You must deliberately carry the attribute-generation process to the point where attributes seem to conflict. That's the way the real design issues emerge.

- In order to reduce the amount of recording of attributes, you may group functions that have similar attributes and apply those attributes to the group. Only the differences then need to be recorded. Be wary, however, of imagining attributes are similar just because you wish there were less to record.

15.6 Summary

Why?
A clear definition of attributes enables designers and their clients to make informed, intelligent decisions on trade-offs. Transforming the attributes in a variety of ways creates new ideas that will contribute to a successful design.

When?
Categorize attributes soon after the functions are first listed, identifying which attributes modify specific functions. When the design process proceeds in levels, repeat the attribute process each time the functions are defined at a finer level of detail.

How?
Follow these steps to refine the set of attributes:

1. Brainstorm a list of possible attributes.

2. Sort attributes from attribute details. Fill in the list with attributes for all the details, and with the details suggested by all the attributes.

3. Assign each attribute to the appropriate function or functions.

4. Classify the attributes into Must (**M**), Want (**W**), and Ignore (**I**).

5. Document the **M** and **W** attributes for further processing.

Who?
The client is ultimately responsible for specifying the desired attributes with guidance from the designers and those people whom the client represents.

16 CONSTRAINTS

Now that all attributes have been clarified, attached to functions, and classified, we are ready to proceed with further clarification of the true requirements. If new requirements and attributes are uncovered by later processes, we simply back up the process temporarily until that requirement or attribute is incorporated into the lists.

To determine what further clarification is needed, suppose that all the functions and attributes have been identified, all the requirements clarified, and a product built. How will the designers know if they've done their job—if the product is actually what was desired? They know the product is finished when all the functions are present—that is, when the function has all its **M**ust attributes (**M** attribute). If not, then *some* function has been implemented, but it's not the function defined by the requirements.

How do you determine if an attribute has been implemented? For an **M** attribute to be present, all its *constraints* must be satisfied. If not, it might be a similar attribute, but it's not the real thing. In this chapter, we'll investigate constraints: what they are, and the processes by which they can be defined.

16.1 Defining Constraints

A *constraint* is a mandatory condition placed on an **M** attribute. In order for the final design solution to be acceptable, every constraint must be satisfied. Therefore, a constraint must be defined in terms that will enable participants to determine objectively whether or not it has been satisfied in the finished product.

Constraints are placed on the **M** attributes as you examine each one on the list. For instance, Superchalk must be manufacturable, which means it must be pure chalk and cylindrical. A cylinder has a length and a diameter, so constraints might be placed on how long or how thick the chalk should be. Purity constraints might be expressed in terms of chemical tests applied to suitable samples.

In the Elevator Information Device, for example, the attribute, "easy-to-select information," may have been applied to the function, "display relevant information to building occupants." We might derive several different constraints on this attribute, as follows:

1. Assistance requests (experienced): This constraint is defined in relation to the average* number of times an experienced elevator rider,[†] using our system for the first time, requires external assistance[††] in satisfying a predefined series of requests for relevant information. To test whether this constraint has been satisfied, we will run an experiment using a random sample of one hundred experienced riders from the mid-Manhattan area. The average number of external information requests per rider must be less than one.

2. Time to access information: This constraint is defined in relation to the average time required by the experienced elevator rider per information query selection in the above protocol. The time to access information must be less than 1.75 seconds.

3. Information access errors: This constraint is defined in relation to the average number of errors made by the experienced elevator rider in executing the above protocol. The information access errors must be less than 0.9.

16.2 Thinking of Constraints as Boundaries

To visualize the meaning of constraints, imagine an n-dimensional space in which *each dimension is one attribute of the requirements*. The general name of this space is the *state space*, because it contains all possible solutions—that is, combinations of attribute details. Of course, visualizing more than three dimensions is rather difficult for most of us, and three is more than some of us can handle. That's why we use two dimensions as a metaphor for n dimensions, which is not much of a problem, since many important design issues can be expressed in two dimensions at a time.

For instance, in Figure 16-1, the length and diameter attributes of a piece of Superchalk are used as the two dimensions of a state space. Within this space, the horizontal and vertical lines indicate the following constraints, which were imposed by physical reality and the capacity of CCC's chalk fabricating machine:

---

*Terms such as this, when used in constraints, must be carefully defined so that there won't be any basis for argument about whether the constraint has indeed been satisfied. We suggest using footnotes for these definitions to keep the sense of the constraint from being lost in the legalities. For instance, "average" might be defined in terms of the exact design of the statistical experiment needed to produce that average.

[†]"Experienced elevator rider" is defined as "an individual who has traveled ten or more floors in self-service elevators at least four hundred times in the past year."

[††]The term "external assistance" refers to the need to consult printed information in the form of prompts or user guides, information that the user is unable to obtain directly from our system but that is required to complete the desired information selection.

a.   the length cannot be negative

b.   the diameter cannot be negative

c.   the length cannot be greater than 12″

d.   the diameter cannot be greater than 3″

The constraint lines represent *bounding lines*, which define a closed (shaded) region called the *solution space.* Any solution must meet all constraints, and therefore any solution must lie within the solution space. In this case, it must have a length greater than zero and less than twelve inches, and a diameter greater than zero and less than three inches. If these were the *only* two constraints in the requirements, then any point in the shaded region could represent an acceptable solution to the Superchalk problem.

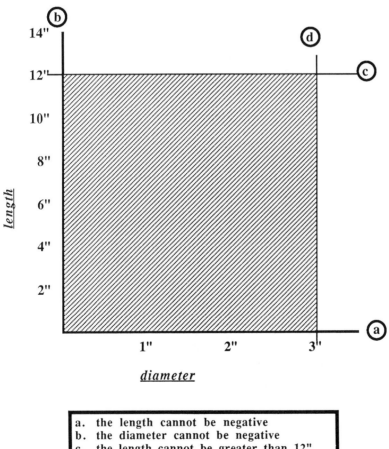

Figure 16-1. The attributes determine the dimensions, the constraints determine the boundaries, and the shaded region represents acceptable solutions.

Figure 16-2 shows the state space concept applied to two attributes of the Elevator Information Device. The dimensions are two of the attributes given above for the function "display relevant information to building occupants":

2.   time to access information
3.   information access errors

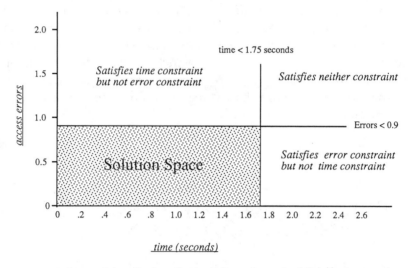

Figure 16-2.   Two attributes of the "display relevant information to building occupants" function. Again, the constraints determine the boundaries.

If there were a third constraint, a third dimension would be involved and the solution space would be a solid figure. Any point within that solid could be an acceptable solution. Of course, we don't know yet if there *are* any achievable points in the solution space—that will be up to the design process to determine. We can see, however, that the tighter the constraints, and the more of them, the smaller the region in which acceptable solutions might be found. That's why we don't want to create more constraints than are really required.

16.3 Testing the Constraints

The solution space model suggests a procedure for testing whether constraints are too strict, or too weak, as described below.

16.3.1 Too strict?

Working with a group of users, conjure up a fantasy solution that falls just outside one of the constraint boundaries. Then ask the group, "Is this solution okay?" For example, the BLT designers might ask Byron, Wilma, and John, "If the chalk is fourteen inches long, is that okay?" If Byron figures out how to adjust the fabricating machine so that the answer is yes, then the length constraint was too strict and the boundary could be moved, thus increasing the size of the solution space (Figure 16-3).

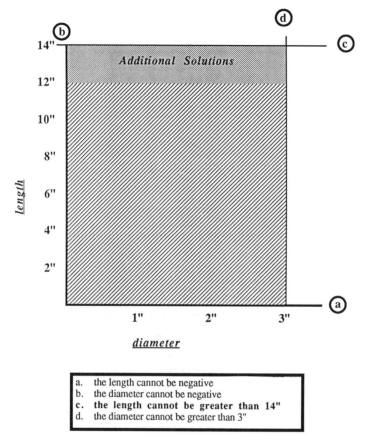

a.	the length cannot be negative
b.	the diameter cannot be negative
c.	**the length cannot be greater than 14"**
d.	the diameter cannot be greater than 3"

Figure 16-3. Relaxing a constraint moves a boundary and increases the size of the solution space and the number of potential solutions.

16.3.2 Not strict enough?

If the answer is no, the constraint is not too strict, but perhaps it is too weak. Next, propose a solution just *inside* the boundary. If this is also okay, then the test passes and the constraint is neither too strong nor too weak. If not, more constraint work is needed.

Suppose you ask, "Is a length of a half inch okay?" John, an experienced chalk user says, "No, because my fingernails will squeak on the blackboard." You keep questioning and eventually establish that one inch is the shortest acceptable chalk length. In a similar fashion, you may determine that diameters less than one-quarter inch are too hard to grip, and are therefore not acceptable. You can now adjust the boundary as shown in Figure 16-4.

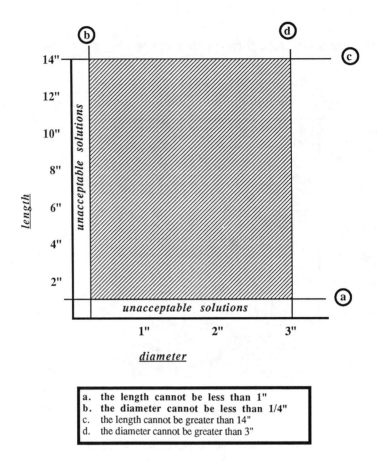

Figure 16-4. Tightening a constraint moves a boundary and decreases the size of the solution space.

16.3.3 Unclear?

Sometimes, this sort of questioning yields neither a yes nor a no. You may hear a maybe, or simply see a lot of puzzled looks, or an argument may flare up between two of the clients. Such responses may indicate that the attribute or constraint definitions are not clear, or in the case of an argument, it may reveal an essential political difference among the clients—a difference they will have to negotiate before the requirements process can be concluded successfully.

16.3.4 Generating new ideas

Merely sketching the solution space will sometimes generate new ideas about constraints, without asking any questions at all. As Byron watched the development of the Superchalk solution space, for example, he got an idea. Putting his finger in the middle of the solution space, he said, "This is where *conventional* chalk is, right? Well, I don't want Superchalk even to look like ordinary chalk." This constraint was represented by putting a "hole" in the solution space, the exact size and shape of the solution space of ordinary chalk. (See Figure 16-5.)

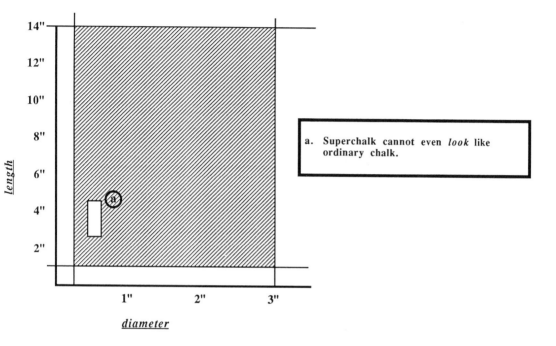

Figure 16-5. Some constraints can put a "hole" in the solution space. This is the solution space of an excluded solution, in this case, ordinary chalk.

Once constraint work is done, do a similar test on points *within* the solution space, but this time ask, "Is this *better* than that?" Such a test reveals and tests *preferences,* as we shall see in Chapter 17.

16.4 Interrelated Constraints

Constraints are sometimes interrelated, which can influence the size of the solution space. If you suspect that this is the case, ask questions involving two constraints at once. For example, BLT Design might ask if length and width are related in any way. Wilma, a professor of materials science, might explain that based on the strength of chalk, a narrow piece cannot be too long or it will break from excessive leverage in normal use. She might supply a formula, which could be used to create the curved boundary in Figure 16-6, thus further reducing the solution space.

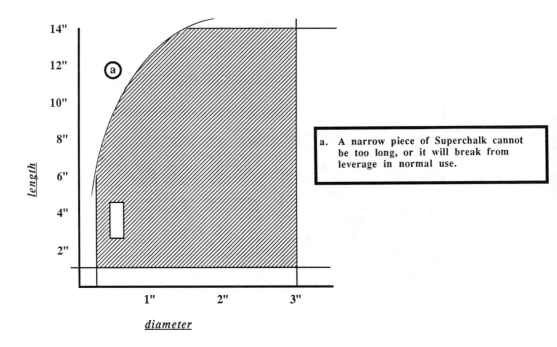

Figure 16-6. Related constraints produce irregular boundaries.

Interrelationships among constraints can also *increase* the size of the solution space. In the elevator information system, we might ask, "What if the response time were under 0.2 seconds. In that case, would it be okay if the number of errors were greater than 0.9, perhaps 1.8, because then the rider could easily ask a different question to correct the error?"

If this reasoning were considered acceptable, then the constraints would be amended to reflect this combined effect, perhaps as follows:

3.   (revised) Information access errors: This constraint is defined in relation to the average number of errors made by the experienced elevator rider in executing the above protocol. The information access errors must be less than 0.9, unless the information access time is less than 0.2 seconds. In that case, the information access errors must be less than 1.8.

Figure 16-7 shows the solution space for these revised constraints. It's a little bigger than the previous space, giving the designers more room to find a truly outstanding solution. Put another way, with these constraints slightly relaxed, some other constraints might be easier to meet.

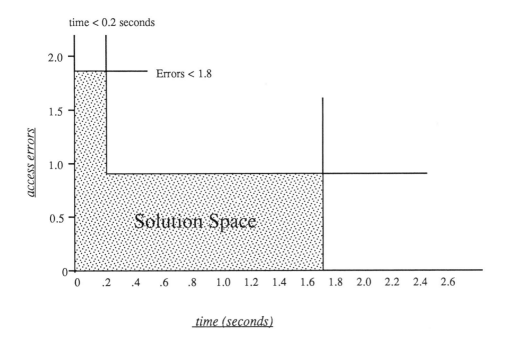

Figure 16-7. When interrelated constraints are relaxed, the solution space grows and becomes irregular.

16.5 Overconstraint

Relaxing a constraint increases the size of the solution space, and therefore may increase the number of potential solutions. Conversely, tightening a constraint decreases the size of the solution space and may eventually result in an *overconstrained* set of requirements. In this event, finding any acceptable solution would be very hard, or even impossible.

It will *certainly* be impossible to find a solution if there are *conflicting constraints,* as when one constraint says Superchalk must be longer than four inches and another says it must be shorter than three inches. If you try to sketch this solution space, you'll find that it has *zero area.*

Even when the solution space doesn't have zero area, it may still be impossible to find any *real* solution possibility that falls within it. Of course, this unfortunate condition won't become apparent until design. Experienced designers, though, can often guess what is happening, especially as they watch the solution space shrinking.

If the solution space is actually void of possible solutions, you will have to negotiate one constraint against another. If you can't do that, you'll have to drop the project. There are worse things than dropping a project in the requirements stage, such as dropping it in the post-implementation stage. So be on the lookout for an empty solution space, and don't be afraid to suggest termination if it's clear that the project is overconstrained.

When and how to do this is very important. Sometimes people are implicitly pre-negotiating constraints, because they have an image of the system becoming overconstrained. For example, they won't mention something they really want, or they'll start trying to suppress someone else's idea.

Einstein once said, "A theory should be as simple as possible, but no simpler." We can paraphrase Einstein and say,

Requirements should be as constrained as possible but no more constrained.

Applying this principle is equivalent to saying, "The solution space should be as large as possible, but no larger." That's why it's a much better idea to deal with constraints explicitly.

16.6 Psychology of Constraints

Mathematically, we know that excessive constraints limit the size of the solution space, but even worse, they limit us *psychologically*.

16.6.1 The tilt concept

If you watch people play pinball, you'll notice that some of them tilt the machine occasionally, while others never tilt. The ones who tilt too much are not good players, because they are unable to restrain themselves; but the ones who never tilt are terrible players, because they restrain themselves excessively. This is the *tilt concept*:

If you never tilt, you're not using your full resources.

As designers, we are often intimidated by constraints. Someone says, "That information is confidential," and we stop pursuing something that may be essential to a good design. Someone points to a seven-foot shelf of standards manuals and declares, "That violates standards," so we drop a promising idea. Someone frowns and says, "The boss would never approve of that," and we shrivel into a ball and change the subject, never daring to check it out.

It's a psychological fact that most of us tend to operate at some distance from the constraint boundaries, because we're afraid of tilting something. Thus, none of us search the entire solution space, but only a reduced solution space, as suggested in Figure 16-8.

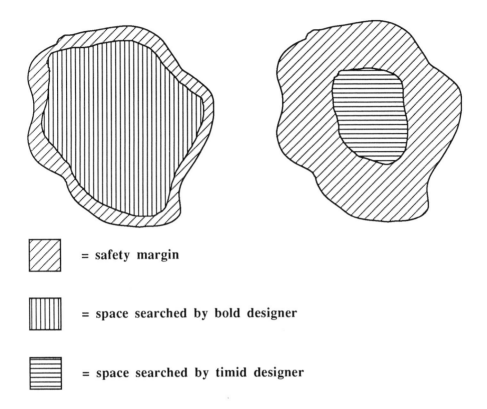

= safety margin

= space searched by bold designer

= space searched by timid designer

Figure 16-8. Most designers don't search the entire solution space, but a reduced solution space determined by their fear of constraints.

16.6.2 Breaking constraints

You can always overcome this limitation by probing politely, but firmly. At one of Jerry's clients, the requirements team indicated that the product *had* to be shipped by November 1. As this seemed an impossible overconstraint, Jerry asked, "Where does that date come from?" Everybody in the room got a shocked and frightened look, and somebody said, "The *boss* told us."

If Jerry didn't understand the tilt concept, he would have become just as frightened as they were, and dropped the subject. But he was an outsider anyway, so he asked for a break and went down the hall to talk to the boss. "There seems to be some uncertainty in the other room," he said. "What deadline did you set on this project, anyway?"

"November 1."

Jerry might have stopped there, too, but the tilt concept kept him going. "That may turn out to be an overconstraint. Can you explain where it comes from?"

The boss looked a little puzzled. "Constraint? I didn't mean it to be a constraint. They asked me when I wanted it, and so I said it would be nice to have it by November 1. I didn't want it *after* November 1 because it would interfere with our year-end processing, but after January 15 would be just fine. Heck, as long as I get it by June there won't be any problem." In other words, there was a hole in the solution space, between November 1 and January 15, but there was plenty of solution space *on the other side of the hole.*

When Jerry returned to the requirements room, the team was shocked to hear the news. "You must be a super negotiator," they said, their eyes sparkling with admiration.

"No, but I'm a terrific pinball player."

16.6.3 The self-esteem bad-design cycle

Now we can see one reason that the quality of the requirements work depends on the personality of the requirements team, and also on the perceived safety of the working environment. A team with low self-esteem will be afraid to check overconstraints, and consequently will create a more difficult design task. Then because the design task is extra difficult, they won't do as good a job as they might have. As a result, their self-esteem will drop, and they'll be even worse on the next project. Only a skilled intervention from outside can break such a vicious cycle of low self-esteem and bad design.

16.7 Constraint Produces Freedom

Let's not give the impression that constraints are bad for design, or even that they're just psychologically bad. A major objective of the requirements process is to document the constraints, all the constraints, and nothing but the constraints. If you overconstrain, you may produce ho-hum solutions, but if you underconstrain, you may produce nonsolutions—products that don't lie within the real solution space. Some people call these "Edsels"!

Furthermore, underconstraint itself can become a constraint. The human mind is limited, so you can never truly search the entire solution space. Therefore, if the solution space is excessively large, you may spend much of your time searching areas that don't contain real solutions. Even if you eventually wind up in the true solution space, you probably haven't had time to search it very well.

16.7.1 Standards

Standards are an excellent example of the freeing effect of certain constraints. If we were designing a bicycle, we'd generally find it rather emancipating to be able to assume standard threading on all bolted parts. The existence of standard threading will also reduce costs, and make it easier to deal with suppliers, so that tightening this constraint loosens up the solution space on several other attribute dimensions.

Certainly there might be one place where a slightly nonstandard thread would produce some sort of benefit, but not having to consider the threading standard on each and every bolt will free us to concentrate on other design issues, and probably result in a better design.

16.7.2 Languages and other tools

In information systems, we commonly see the constraint of underconstraint when programmers decide that none of the thousand or so existing programming languages is quite right for their project. So, as a preliminary step, they set about producing a programming language, rather than a program. Two years later, they are still working on their "preliminary" step. It's just barely possible that one of the existing programming languages could have been used to produce something rather useful and not totally inelegant in less than two years. But as one programmer put it, "I would rather write programs that write programs than write programs."

Perhaps it is appropriate to conclude this chapter by mentioning that the same dynamic commonly takes place when members of a product development team decide they need a new methodology. After a few years, their management decides to find out why there's no product, only to discover that their motto has become, "We would rather develop products that develop products than develop products." Many projects—many *companies*—have died with this motto on their walls.

16.8 Helpful Hints and Variations

- The ultimate definition of each constraint must be complete, consistent, and precise enough to settle (or prevent) any argument about whether or not that constraint has been satisfied in the finished product. Such precision is not practical to obtain with many people. At the early stages of the requirements process, you will want to identify all constraints, as well as some of the major concerns addressed by that constraint. Later, the task of drafting constraints can be broken into subtasks, each assigned to a small, qualified group with access to appropriate reference material. When their drafts have been completed, they are then brought back to a larger, representative body for technical review.

- If you are having trouble defining a constraint, remember that the constraint is always answerable by "yes" or "no" or "we can't say 'yes' or 'no' at this moment, but this is exactly what we have to do to find out."

- In order to be able to answer the yes-no question about a constraint, it must be free of all qualitative words.* Suppose we were designing a system and somebody said, "The data transmission rate must be *high*." If the data trans-

---

*For a list of dangerous words to look for in requirements, see Daniel P. Freedman and Gerald M. Weinberg, *Handbook of Walkthroughs, Inspections, and Technical Reviews*, 3rd ed. (Chicago: Scott, Foresman and Co., 1983).

mission rate were 1,200 bits per second, would this constraint be satisfied? To some people, 1,200 bits per second is "high," while to others it is "low." A better statement would be, "The data transmission rate must be *greater than 6 million bits per second.*" That might not be the right constraint, but at least there's no doubt of the answer. Better to be the starting point for an argument than the ending point for a project.

- Although every constraint must be measurable, sometimes the group gets hung up over measurability. An effective way to resolve such tangles and get the requirements work moving forward is to ask, "Can we agree on a person or group we would trust to resolve this issue by developing a way to measure this constraint?" Sometimes, the group that's established consists of the *only* two people who are arguing. "If they can agree," said one participant, "I know I'll be satisfied."

- Another way of resolving measurability conflicts is simply to define *measurement by authority.* "What person or group can be given the authority to say 'yes' or 'no' about the satisfaction of this constraint?" Sometimes there is an official agency: "If the federal bank examiners will accept the auditability of this system, then we will be satisfied that the auditability constraint has been met." Other times there is an official *process:* "If the standards committee conducts a technical review and certifies this system, then the standards constraint will be met."

  Of course, any measurement approach that requires referral to a group of people could slow the design work. When you actually get to the design, it's often possible to speed up the actual measurement process by some practical approximation.

- Suggesting the possibility of dropping the project can have a salutary effect on negotiations to relax constraints, but never threaten. Just point out that having no project at all is one alternative, then ask the group if that alternative is worse than relaxing some hard-and-fast constraint.

- In some organizations, the issue of standards lays like a dead hand on the requirements process. Every time you wish to consider relaxing some proposed constraint, the standards shroud is brought out to bury the subject. One way to nullify this effect is by suggesting a fantasy experiment: "Let's *pretend* we didn't have a standard on that, and just brainstorm what might happen, good and bad." As long as you convince the group that it's just a fantasy, they can loosen up and get an idea if this standard is a constraint whose relaxation is worth pursuing. After all, standards were written by people, and they can be changed by people.

- Use a similar process in those situations that threaten to be underconstrained, such as when participants in every new programming project believe it requires the design of a new language, or the creation of a new software development methodology. Suggest a similar fantasy experiment: "Let's *pre-*

tend we used the same old boring language and the same old boring method-ology, and just brainstorm what might happen, good and bad." Again, as long as you convince the group that it's merely a fantasy, they can loosen up and perhaps see that there is sufficient challenge in the product itself not to require spicing it up with doses of tool building or meta-development.

16.9 Summary

Why?
Try to develop constraints explicitly, because they become the approval/disapproval criteria of the system being built. Nobody can look at an attribute and know on sight whether it meets a constraint or not. Only the client can identify the con-straints, and one person's constraint might be a matter of total indifference to another.

When?
Develop constraints only after the attributes have been fully developed and classi-fied, but before attempting to move on to other requirements steps.

How?
When developing a list of constraints, follow this process:

1.   Generate a list of constraints based on the **M** attributes.

2.   Test the constraints list for correctness and completeness.

3.   Look for interrelated constraints that may produce either a smaller or larger potential solution space.

4.   Test carefully for overconstraint, by exploring both just inside and just outside the constraint boundaries.

5.   Negotiate as necessary to get as large a solution space as possible, but no larger.

Who?
Developing constraints involves all the clients who have the right of approval or disapproval of the finished system, because the constraints ultimately come from them. Sometimes the designers participate, if they are concerned about overcon-straint.

17   PREFERENCES

Barbara, Larry, and Todd are preparing for their next meeting in the chalk cave at CCC. "Although we're not quite finished with requirements," Barbara says, "it's probably a good idea if we go to the meeting prepared with some design ideas. . . ." She pauses, smiles a little smile, and then continues, ". . .to chalk up some points with Byron."

Todd and Larry don't respond, so she goes on, "To summarize what we know so far, I've sketched the solution space we developed in the previous meeting (Figure 17-1). Points A and B are two possible designs for Superchalk that I thought were interesting. What do you guys think?"

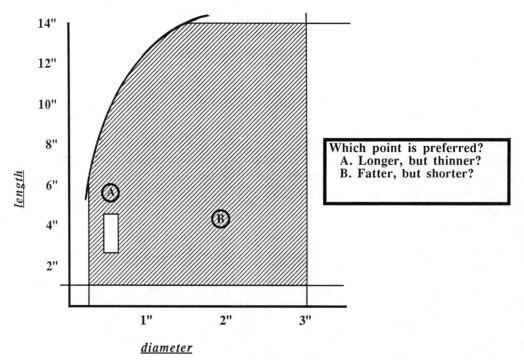

Which point is preferred?
A. Longer, but thinner?
B. Fatter, but shorter?

Figure 17-1. The solution space for Superchalk, with two possible designs indicated by points A and B.

> Reader: Before reading further, try to decide which design point in Figure 17-1, A or B, is preferable for Superchalk, and why.

"I think that A is better," says Todd. "It's thinner, so it will use less material, which will lower the cost to manufacture."

"No, I think B is better," counters Larry. "The thicker chalk makes it look more 'super,' so it should sell better than the competition. That means more profits."

"No way," says Todd. "Profit is sales *minus* costs. Any additional sales you *might* get will be eaten up by the excess costs."

"I don't think you understand the chalk business," Larry says. "For one thing, packaging is a major part of the cost equation, otherwise you lose too much in breakage. So a little more material is a minor part of the cost, and the extra material will mean less breakage, which means fewer returns and more profits."

Todd looked toward Barbara, frustration on his face. "You're the senior designer, Barb. What do you think?"

"I think," she said, "that you two need another lesson in requirements. You need to learn about *preferences*."

"What about preferences?"

"Rather than have me explain it," she said, "here's a new book by Gause and Weinberg. Read Chapter 17 before you come to our next meeting with Byron."

17.1 Defining a Preference

A *preference* is a desirable but *optional* condition placed on an attribute. Any final design solution that satisfies every constraint is an acceptable solution, but some acceptable solutions might be preferable to others. Preferences enable the designer to compare acceptable solutions and choose the better ones.

As designers, we are unwilling to pay even one additional penny for an improvement in our solution with respect to the constraints, as long as we have met the specified levels. We'll take all of the preferences we can get, however, as long as we don't have to pay extra for them.

17.1.1 An example

For example, suppose the client for our Elevator Information Device has decided that "flexibility of information presentation" is an attribute to be considered a preference. We would need to state this preference in a manner that is consistent with the client's wishes, and that expresses the relationship between the quantity of the attribute and its desirability. For example,

•   flexibility of information presentation: The system is to provide maximal form and format control on all information presentation, so that the local building management, in consultation with frequent elevator users, can determine and adjust the form in which information is to be transferred to the rider. This feature is to make it possible for the Elevator Information Device to adapt, through an appropriate control individual, to local needs, custom, and culture.

Without this definition to clarify the attribute, some people might think that flexibility meant that *any* individual could change formats for presentation of information. The definition makes clear that the client neither desires nor expects this kind of flexibility, but only the flexibility of having a specialist change the presentation for all users on one elevator.

17.1.2 *The origin of preferences*

Now we can consider the question that Barbara asked Todd and Larry. Which Super-chalk is preferable, A or B? The right answer, at this point in the requirements process is *we don't know.* Indeed, when they got to the meeting, Byron explained that he would prefer a third point in the solution space, somewhat above both of the others and between them.

"It would be a nice marketing feature if the chalk fit in standard chalk holders," he explained, "but was sufficiently longer than standard chalk to make people think they were getting something 'super.' "

The moral of this story is quite simple, and one that designers should never forget:

Preferences come from the client, not the designer.

It's easy to forget, especially because the clients may not be aware of their preferences unless and until the designers offer them options.

17.2 Making Preferences Measurable

Preferences are used by designers to guide them in satisfying their clients. Therefore, preferences won't be of much use unless each is defined in terms that will enable designers to determine to what degree the preference has been satisfied.

17.2.1 *A reasonable approach to metrics*

Much has been written about the importance of metrics—units to measure the amount of some attribute—but don't get bogged down over metrics. Remember, as we've said before, the actual metrics are nothing; the process of "metrifying" is everything. What's important is making the effort to identify a metric for every preference, because the thought process is going to help everybody understand exactly what the preference is.

Make the attempt to define a metric, but ease off if you've been trying for a while and getting stuck. First, check whether there are actually two or more preferences mixed into one, which will always confuse attempts at metrification. Write each of the proposed metrics on a separate flip chart, then ask for each, "What is the name of the preference that this metric measures?" This process will usually yield similar, but distinct, names, revealing the existence of similar, but multiple, preferences. If that doesn't resolve the problem, set up a separate task group to define the metric at a later time, then proceed to the next preference.

17.2.2 Making the preference measurable

The previous definition of "flexibility of information presentation" falls short of defining how flexibility will be measured. Consequently, it is not a complete definition, and so the designers will have trouble using it to choose among points in the solution space. The following definition of another preference shows more attention to measurability:

- polysensory information presentation: The system is to provide as much information as possible in a form that uses more than one sense in a range of each sense that is effective for each rider. The reason for this redundancy of senses is to provide better information transfer in the presence of noise, backup in case of failure of one form of presentation, as well as providing information to people who lack some or all capacity in one or more senses. This attribute will be measured by the percentage of people in a random sample who can receive one hundred percent of the information they require from the system on at least two sensory channels.*

Note that at this early stage, you may be unable to furnish the precision and measurability that will ultimately be needed. Indeed, you do not want to burden the requirements process with excessive precision. Nevertheless, any efforts to get started in this direction will invariably pay huge dividends in satisfaction with the ultimate design. Measurable attributes will allow the designer to explore the solution space in such a way as to maximize the *flexibility of information presentation* and *polysensory information presentation* preferences, while meeting all the constraints.

17.3 Distinguishing Between Constraints and Preferences

Although constraints are attributes the solution must have, preferences are attributes that determine conditions the clients would like to improve the solution. To illustrate the difference between these two concepts, here's a story: Once upon

*The measurability of this preference could be further improved by describing the process of sampling and testing in more detail. On the other hand, too much precision too early may impede the requirements work. Give the overall idea of the measurement in one paragraph, then footnote the precise details of the measurement technique.

a time, Jerry's New York lawyer, Herbert, drove Jerry uptown to a meeting to settle some contracts. When they arrived at the meeting place, the lawyer parked right next to a sign reading, "ABSOLUTELY NO PARKING." When Jerry pointed out the sign, Herbert shrugged and remarked, "I may get a ticket, or I may not. The chance is fifty-fifty, and the fine is $50. So, I just think of this as a parking space that costs $25. At my rates, neither you nor I can afford to look for a cheaper place."

To Jerry, "no parking" was a constraint: He must not park there, whatever the cost. To Herbert, there was only a preference: Try to minimize total costs of the trip, including parking and his fees. Nothing about an attribute itself tells us whether it is a constraint or a preference.

Only the strength of the client's fears or desires determines which is a constraint and which is a preference.

17.3.1 Is meeting the schedule a constraint?

To show the importance of this distinction between constraints and preferences, take the common and often devastating example we discussed in the previous chapter, getting the system built on schedule. To the team members, November 1 was a constraint, while to their boss it was only a preference, and a mild one at that.

It may shock some people to learn that in some projects meeting the schedule may be a constraint, but in others it may be a preference. And sometimes it is neither. Let's see what consequences ensue when we make the wrong assignment.

An information system designed to tabulate and report U.S. presidential election results *must* be ready to operate by the first Tuesday after the first Monday in November of the next leap year. If not, it has no value whatsoever until four years later. In this case, meeting the schedule is clearly a constraint.

An information system designed to provide management reports on monthly sales could be nice to have by next month, but may not be *required* by next month. If we've been operating our business successfully without this information for thirty years, waiting another month or two may not be a life-or-death matter, though some people may act that way.

Many development projects suffer horribly from the failure to distinguish whether meeting the schedule is a constraint or a preference. People rant and rave as if the schedule is a constraint, and often generate poor quality in their quest for speed. When the poor quality finally faces the realities of testing, the project slows down, the schedule is not met, and the project is delivered "late." Then they discover that meeting the schedule was not really a constraint at all, because the project is accepted, "late" and all.

An information system designed to automate data acquisition in a university research laboratory may have meeting the schedule as neither a constraint nor a preference. A graduate student is assigned the job, and it will get done whenever it gets done. If it doesn't get done at all, then it's all written off as graduate education. Data keeps on being acquired in the old way until some other graduate student takes on the job and manages to finish it.

Many programmers and engineers who were trained in relatively casual university environments have difficulty distinguishing between schedule constraints and preferences, or even mere frills. This difference in training accounts for a good deal of the frustration of managers who hire computer science and engineering graduates—frustration that often leads to mistakenly imposing every schedule as a constraint. This frustration could be alleviated if those managers carefully identified and then communicated the differences between constraints, preferences, and frills.

17.4 Constrained Preferences

Constraints enable you to distinguish solutions from nonsolutions. You either satisfy the constraint or you have no solution. Preferences, on the other hand, are the things you can't get enough of, and therein lies the great danger of preferences: *greed.*

If there's one thing worse than setting an inappropriate time constraint, it's setting an inappropriate time *preference.* If you know, for example, that you must produce a system by July 1, you make plans and get to work. But suppose you're told you must produce a system *as soon as possible.* How will you know if you're ever doing enough? If you finish by June 1, perhaps there was a way you could have finished by May 15. . . or May 8. . . or May 7.

Consultants can always recognize projects that take place in an environment of unconstrained preferences: Just notice the climate of *unrelenting panic.* If you want to stay out of panic mode, at least some of the time, you must learn to put constraints on your client's preferences. Here's how this could be done with one of our example preferences:

- polysensory information presentation (constrained): The system is to provide as much information as possible in a form that uses more than one sense in a range of each sense that is effective for each rider. The reason for this redundancy of senses is to provide better information transfer in the presence of noise, backup in case of failure of one form of presentation, as well as providing information to people who lack some or all input capacity in one or more senses. This attribute will be measured by the percentage of people in a random sample who can receive one hundred percent of the information they require from the system on at least two sensory channels.

  We require that the system be polysensory for at least thirty percent of people in the sample. We would like to do better, but we aren't interested at all in more than ninety percent.

A good way to discuss constraining objectives with your client is by using a sketch of the solution space. As you indicate various points in the solution space, ask the clients to discuss their feelings about being at that point. By listening to their feelings, you can usually shade out a region of the solution space that the client really isn't willing to die fighting for. Figure 17-2 shows how BLT Design might shade the Superchalk solution space after a short discussion with Byron.

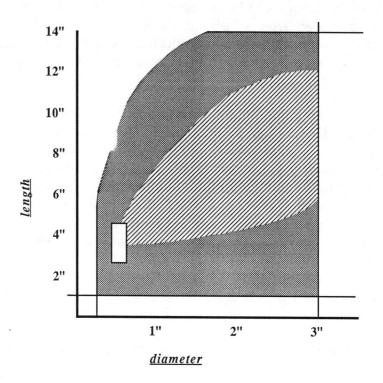

Figure 17-2. The darker shaded region of the solution space indicates preferences that aren't really that important, as derived from discussion points in the solution space with the client.

17.4.1 What's-it-worth? graphs

An even better way to constrain preferences is to provide a function describing the preference levels versus the payoff for reaching those levels. We call this the *what's-it-worth? function*, or in graphical form the *what's-it-worth? graph*. Figure 17-3 shows two possible such graphs for the statement, "We require that the system be poly-sensory for at least thirty percent of people in the sample. We would like to do better, but we aren't interested at all in more than ninety percent."

Even though the two lines in Figure 17-3 are consistent with the same verbal statement, the graph reveals them to be rather different. Preference #1 says that at thirty percent, the polysensory attribute suddenly becomes of value, but improvements over that level are not worth too much. Preference #2 says that the value of polysensory information presentation is small at thirty percent, but rises rapidly thereafter.

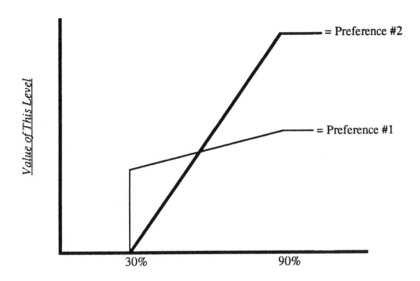

Polysensory Information Presentation

Figure 17-3. The what's-it-worth? graph reveals the ambiguity in the verbal statement of the constrained polysensory information presentation preference. Both lines are consistent with the verbal statement, "We require that the system be polysensory for at least thirty percent of people in the sample. We would like to do better, but we aren't interested in more than ninety percent."

Designing for preference #1 would have a rather different emphasis than designing for preference #2. Without the what's-it-worth? graph to clarify the client's valuation of this preference, the designer would be forced to guess. Since these two curves are only two out of thousands of possible meanings, the chances of guessing right are not very awesome.

17.4.2 When-do-you-need-it? graphs

An especially important kind of what's-it-worth? graph involves the interpretation of schedules, producing a *when-do-you-need-it? graph*. Figures 17-4 and 17-5 show two possible what's-it-worth? graphs for a schedule preference stated "as soon after July 1 as possible." Both graphs represent a situation in which the system cannot be used before July 1, so they show zero value for delivery before that date.

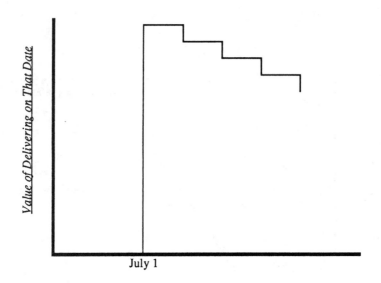

Figure 17-4. The what's-it-worth? graph of a delivery date preference of "as soon after July 1 as possible."

The situation in Figure 17-4, however, might be represented in words as follows: "We won't be ready for it until July 1. If we get it then, we'll possibly be able to use it a little, but if we get it later, it will still be valuable. We can't wait forever, though, because each month that passes costs us the value of using it for a month."

Figure 17-5, on the other hand, might be saying, "We really must have it by September 1, otherwise it will have no value to us. We won't be ready to receive it until July 1, but if we do get it then, we'll get a chance to practice with it before September 1, and that will be of some value to us." Can you see how the two projects would unfold very differently, even if all that differed was this graph?

17.5 Reframing Constraints into Preferences

The what's-it-worth? graph is an antidote for greed. Without it, designers may strive to get as much as they can of one preference at the sacrifice of all the others. Carried to an extreme, this pursuit of a single preference produces the dreaded designer's disease, *optimitis*.*

---

*For a more complete discussion of optimitis and its cure, see Gerald M. Weinberg, *Rethinking Systems Analysis & Design* (New York: Dorset House Publishing, 1988), pp. 111ff. Another source is Weinberg's *The Secrets of Consulting* (New York: Dorset House Publishing, 1985), pp. 22-28.

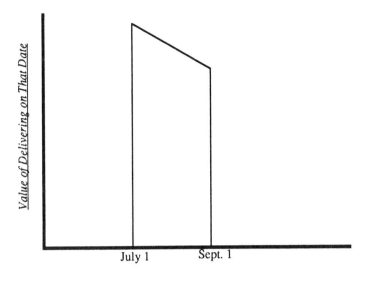

Figure 17-5. The what's-it-worth? graph of a delivery date preference of "as soon after July 1 as possible," but with completely different consequences.

17.5.1 Trading off among preferences

Curing optimitis requires tools, such as trade-off charts and what's-it-worth? graphs, to remind people that in design there is no free lunch. Getting more of one preference may seem temporarily free, but eventually it starts to hurt other preferences. At the very least, operating near the boundaries of the solution space tends to reduce the flexibility of a design.

High-performance racing cars are an excellent example of trade-offs. Usually, no more than a third of the cars at the Indianapolis Motor Speedway finish the race, because the quest for optimal speed incurs a cost in reliability. At the Speedway, of course, sacrificing even two miles per hour may guarantee that you can't win the grand prize, so the what's-it-worth? graph for speed takes a nosedive below, say, the qualifying speed of last year's winner. Another sharp drop in the graph appears at the speed of last year's slowest qualifier. There's some value to just being in such a prestigious race, but if your car doesn't even qualify, it probably costs you a lot more than you get out of it (see Figure 17-6).

It's evident from Figure 17-6 why racing car designers are willing to trade their reliability preferences, if they must, in order to get above a critical lap speed. But in order to move to higher speeds, they might have to use newer, riskier technologies. Figure 17-7 shows the trade-off that might face a crew whose car failed to qualify last year. By introducing a new fuel injection system, they feel they can boost their speed, but with a certain risk of failure. This risk, however, would put them into contention for qualifying, so their payoff would be better.

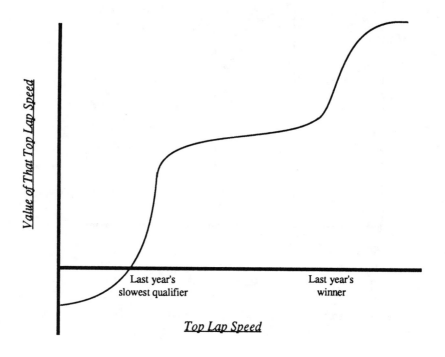

Figure 17-6. The what's-it-worth? graph for top lap speed for a car at the Indianapolis Motor Speedway.

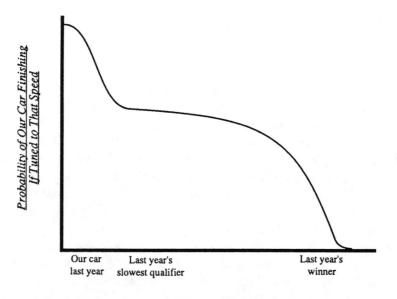

Figure 17-7. The probability of finishing for a car at the Indianapolis Motor Speedway, based on the designed top lap speed.

But as Figure 17-7 also shows, the new fuel injection system won't boost them into the winning range. Should they then risk installing the new crankshaft, which might work, but has a very high chance of failure? The decision depends on the perceived values of the various outcomes.

Figure 17-8 shows two tables of values assigned to various outcomes: winning; not winning, but placing high; simply qualifying to be in the race, but having no real chance of placing high; and not qualifying at all. In case A, winning has such a high value that the designer would be willing to risk putting in the new crankshaft, even though there's a high chance the car won't even finish. In case B, however, winning isn't quite as valuable, making the new crankshaft a bad bet.

	Speed	Perceived value	Prob. of qualifying and finishing	Expected value of the race	Expected net value
Last year's non-qualifying	180	0	0	0	-200
Should qualify but not win	182	100	.75	75	-125
Should place in a high position	184	500	.60	300	+100
A finish should win the race	186	10000	.20	2000	+1800

Case A. The owner places a high (10,000) value on actually winning, and thus would be willing to risk installing new equipment that might reduce chances of finishing.

	Speed	Perceived value	Prob. of qualifying and finishing	Expected value of the race	Expected net value
Last year's non-qualifying	180	0	0	0	-200
Should qualify but not win	182	100	.75	75	-125
Should place in a high position	184	500	.60	300	+100
A finish should win the race	186	1000	.20	200	0

Case B. The owner places a low (1,000) value on actually winning, and thus would be less inclined to risk not finishing by installing new equipment.

Figure 17-8. Two tables showing how the value assigned to winning affects design decisions for a car at the Indianapolis Motor Speedway.

In racing, sometimes a car owner puts such value on winning that nothing less than winning is perceived to have any value. Winning then becomes a constraint—"winning at all cost"—and none of this kind of trade-off analysis makes any sense. But if the constraint is transformed into a preference to maximize profits from the race, you'll want to constrain the other preferences, such as top speed. In other words, all requirements ultimately depend on the client's value system, which the requirements process had better make totally visible.

17.5.2 Zeroth Law of Product Development

You can now see why it's so important to distinguish between constraints and preferences. Preferences only begin to have meaning when all the function-attribute-constraints have been satisfied—in other words, *when we are already in the solution space*. Preferences indicate which solutions are preferable within the solution space. But if winning was a constraint, then it's no consolation to know that our car had the fastest qualifying time, or led for the most laps, or had the most sponsors.

It's a lot easier to design a race car if it doesn't have to win. This is the *Zeroth Law of Product Development:*

**If it doesn't have to satisfy the constraints,
then it can satisfy any other requirement.**

If a computer program doesn't have to perform the functions that the clients say it must have, then the program can be coded in any language, can be made to run very fast, and can be built at a very low cost. In fact, since it doesn't satisfy the constraints, then from the clients' point of view it's really doing nothing. Thus, it could just as well be a single no-op instruction—it's fast, cheap, and works in just about any language.

The Zeroth Law is not, of course, confined to software development, but applies universally. You must never let preferences slip over into becoming constraints that aren't really constraints, or vice versa!

17.6 Helpful Hints and Variations

* As with constraints, the ultimate definition of each preference must be complete, consistent, and precise enough to settle (or prevent) any argument over to what extent that preference has been satisfied in the finished product. As we know, this kind of precision is not practical to obtain in a large room filled with many people. Therefore, at the early stages of the requirements process, try to identify every preference, as well as some of the major concerns addressed by that preference; and later, the task of defining preferences can be broken into subtasks, each assigned to a small, qualified group with access to appropriate reference material. When the drafts for each preference

have been completed, they are then brought back to a larger, representative body for technical review.

- Many design projects establish constraints so tightly that the implicit preference is "find any solution that meets the constraints, then breathe a sigh of relief and quit." Learn to listen for pressure for premature closure of defining preferences as a symptom of overconstraint. What is premature? Well, if the project is going to take a year, for example, a day or two devoted to preferences is reasonable.

- Warning to clients: You may be tempted to say that all attributes are preferences. Then because there's no free lunch, designers will make decisions based on their own hidden preferences, hidden even from themselves. The worst case is where the designer tells you, "No problem!" then makes decisions for you. You can protect yourself from this kind of underhanded designer by applying the "Orange Juice Test,"* a method for choosing services derived from a method for choosing a hotel for a conference. The method is based on listening to responses to an unreasonable demand, like having seven hundred large glasses of freshly squeezed orange juice for a 7 A.M. breakfast. If the hotel people say, "No problem," you know they're going to disappoint you.

17.7 Summary

Why?
Constraints define the admissible region of solution space, while preferences guide the search through the admissible region. Without preferences, designers could stop at the first acceptable solution (any one that meets all constraints), because they have no guidance as to what the clients will consider "better."

When?
Preferences are developed after constraints have bounded the solution space, although since the process of developing preferences generally leads to some modification of constraints and vice versa, the process is more of a cycle.

How?
Follow this cycle:

1. Develop a broad list of preferences.

2. Try to convert each preference to a measurable preference, so the designers will know exactly how to measure when they are doing better, and when worse. However, keep in mind what von Neumann said, "There's no sense being precise about something if you

---

*For an explanation of the Orange Juice Test, see Gerald M. Weinberg, *The Secrets of Consulting* (New York: Dorset House Publishing, 1985), pp. 32ff.

don't know what you're talking about," and don't get bogged down in metrics.

3. Reconsider your list of constraints to see if some of them are really preferences. Whenever possible, always reduce constraints to preferences, perhaps constrained preferences, to give the designers a wider solution space to search.

4. To aid in setting preferences clearly, develop what's-it-worth? graphs, which help resolve verbal ambiguities, especially confusion between constraints and preferences.

Who?

Anyone who will be involved in judging the value of the final solution should be involved in the process of setting preferences. If not, these forgotten people will appear in the final act and create a surprise ending by introducing a "new" preference.

18 EXPECTATIONS

If you board the elevator in the morning and ask it to provide a glass of freshly squeezed orange juice, will you be disappointed if it provides frozen juice? Or will you be delighted that it provides juice at all? If you buy a box of Superchalk and find that it doesn't fit in your chalk holder, will you be disappointed? Or will you be delighted that its fatter shape makes it easier to grip, and renders the chalk holder superfluous?

As every experienced designer knows, the difference between disappointment and delight is not a matter of delivery, but of *how well delivery matches the clients' expectations.* If we think of the design process as a way of providing delight and avoiding disappointment, we'll never be successful if we can't control expectations. That's what this chapter is about: first understanding why different expectations arise, and then applying a heuristic for discovering and limiting the clients' expectations. This *expectation limitation process* is used to explore and clarify users' expectations and is performed for each user constituency.

18.1 Reasons to Limit Expectations

If all the prior steps have been done properly, the expectation limitation process would be logically redundant. As it is, an additional step is needed to limit expectations because

1. People are not perfect.

2. People are not logical.

3. People perceive things differently.

4. Designers are people, too.

Let's look briefly at each of these factors, as a prologue to explaining the expectation limitation process.

18.1.1 People are not perfect

People are not perfect, and we can assure you that in our experience, nobody ever really does everything right up to this point. (We hope this is reassuring, not discouraging.) The expectation limitation process, among other things, acts as a check on the imperfections of earlier stages in the requirements process. Having such a step near the end also serves to demonstrate to participants that you don't expect them to be perfect. This in itself should limit their expectations in a healthy way.

18.1.2 People are not logical

As *Star Trek*'s Mr. Spock always says, "Humans are not logical," and this includes your authors. Some years ago, Jerry clipped a coupon for a free night's stay at any Holiday Inn with one night paid. In order to save the money, he carefully chose a Holiday Inn over a much nicer Ramada, but when he checked in and presented the coupon, the desk clerk wouldn't accept it. "But," Jerry objected, "it says it's valid at *all* Holiday Inns."

"No it doesn't," said the clerk triumphantly. "It says all *participating* Holiday Inns. *This* Holiday Inn isn't participating in this promotion." The Ramada across the street wasn't participating either, but Jerry decided to move there anyway. Still, Don thinks it was unfair of Jerry to punish Holiday Inn because he couldn't read plain English. Even so, Jerry's negative reaction is rather typical of customers who discover upon delivery the true meaning of some "invisible" word in the requirements document. Like Jerry, they will suspect that they have been intentionally misled.

Even though we imply that the requirements document states all and only what the system will do, people may have unstated expectations about things that are not mentioned in the documents, or never discussed in the process. We have many examples in everyday life—warranties and insurance policies come immediately to mind. Our reaction to them demonstrates that being logically, or even legally, correct is not the same as satisfying the user.

18.1.3 People perceive things differently

One of the most interesting systems development projects we ever experienced was the creation of the worldwide space-tracking network for NASA. Each station was designed from the ground up, and, since many tracking sites were in remote and often primitive areas, that really meant *ground* up. Many stations had no local power supply adequate to operate the requisite radars and computers, so stand-alone diesel power was provided.

As it turned out, however, the power supplies were designed by experienced radar people because at that time the computer people had little or no experience in this kind of stand-alone power. Once the station went into operation, the computers crashed every time one of the spare diesel generators went on- or off-line. The problem was traced to electrical transients. The computer people pointed haughtily to the requirements document that said, in effect, "There shall be no transients that affect the continuous operation of the electronic equipment."

But the radar people argued that the diesel power system they had designed met that requirement. "It doesn't disturb the radars at all," they said. "In fact, we overdesigned it a bit, so it shouldn't cause you any trouble." The problem, of course, arose from the two cultures of radars and computers, each of which used similar language to mean rather different things.

As it was too late to redesign or replace the diesel power, a signaling system was installed so that the power engineer could warn the computer operators when a switchover was about to occur. The computers were then shut down and a signal sent back to the power station that it was okay to switch generators—a clumsy but operable solution.

One day, the computer operators were going out to lunch and decided to shut down the computer equipment in case the power was switched over while they were out. As they left the computer building and started down the hill, the power engineer raced out of his building, screaming, "What the #*@!*%# are you trying to do! You #*@!*%# nearly cracked one of my flywheels!"

"What do you mean?" the computer operators asked in all innocence. "We didn't do anything."

"Well, somebody cut the load in half," he screamed over the roar of the diesels. "My engines almost tore themselves off their blocks. If that happens again, I could be killed."

Once again, the problem proved to be a misunderstanding between the two cultures. When the radar equipment shut down, its power requirement declined in stages. When the computer system went off, it shut down all at once. As simple an act as shutting off the power was entirely different for the two types of equipment and could have cost someone's life.

This problem was solved with the signaling system, too, but catching and fixing these problems in the requirements process would have been much better. Eventually, all differences in expectations get worked out, one way or another, but it's obviously a lot easier to use an explicit, up-front process to catch these differences.

In requirements work, you want to express information in limitation form, which is the easiest way for customers to understand special details. It's your job to help the customer understand, in contrast to those whose job it is to write insurance policies, warranties, and phrases like "participating Holiday Inns." These people are trying to protect their own interests—but not trying to have customers notice the exceptional cases. If they *were* trying to help customers notice, they could write things differently, for example,

WARNING: Not all Holiday Inns are participating in this offer. Be sure to check with each Inn at the time you make your reservation.

18.1.4 Designers are people, too

Designers, of course, have all the above human characteristics, plus some others as well. For instance, designers, even more than ordinary mortals, tend to be enthusiasts for their own ideas and oblivious or resistant to the ideas of others. To balance their enthusiasm, they need a warning system, so that when they begin to be carried away by some great idea, they must stop and consider limitations. This extra step acts as a damper on unbridled egoistic enthusiasm.

18.2 Applying the Expectation Limitation Process

To counter the effects of the above all-too-human factors, we've developed a heuristic for exploring and clarifying user expectations, called the *expectation limitation process.* It consists of four steps.

18.2.1 Generate a specific expectation list

First, start with a comprehensive list of user expectations, which may be obtained in a face-to-face session of representative users, or by some sort of user survey,* or a little of both. For face-to-face work, we like to trigger the users with an open-ended question, such as

> "What are you most looking forward to in this system?"

> "Which part of the new system will be most valuable to you?"

The answers become the expectation list (see Figure 18-1).
   Interviews are essential, because you have to listen very carefully to uncover general expectations that are only implied in specific examples. For instance, one user might say, "When I ride the elevator, I can find out the important stuff that I haven't had time to read in the newspaper." In order to understand this expectation, we'll have to ask just what he means by "important stuff" in the newspaper. We might be surprised to discover that the only parts of the paper he reads are the stock quotes and the fashion pages because he is in the women's clothing business.

18.2.2 The elevator example

The generated list of specific expectations for the Elevator Information Device might include items such as these:

---

*In Chapter 21, we will develop a user satisfaction test for surveying users throughout the development process, and after. Results from this test can indicate disappointed (and hence previously unidentified) expectations; follow-up interviews will also augment the expectation list.

Figure 18-1. "What's the thing you're most looking forward to in this product?" From the same piece of Superchalk, some people will want the world. Others will want to save time, to be loved, to get justice, to win friends, to get a raise, to help their country, or to win recognition. And some won't know what they want. In any case, the question will reveal user expectations, and lead to clarifying the product's limitations.

1. "When I ride the elevator, I'll be able to find out the important stuff that I haven't had time to read in the newspaper."

2. "I never have time to take my blood pressure. Riding the elevator is a waste of time, and I'm trapped there so I might as well have my blood pressure taken."

3. "I'll second that, and I want to know what the blood pressure means. I need to know if I have to slow down."

4. "All that stuff is okay, but crime is the big problem in this building. It's going to be harder to get tenants if all this mugging and robbery continues. To me, the most important feature is to recognize criminals and convince them that they won't get away with anything in our building."

5. "I don't want any more breakdowns. This device has got to give plenty of warning to the superintendent when something's not right with the elevator."

6. "It's got to be quieter than the elevators we have now. I can hear people arguing in the elevator, and it disturbs my patients."

7. "Also, by the time my patients get finished riding up that elevator, they are nervous wrecks. I spend the first twenty minutes just calming them down from the elevator ride. Somehow, I expect this device to keep them calm, or at least distracted."

8. "You know, warning me that the elevator is about to fail is very nice, but my experience is that some of these warning devices break down more often than the elevator itself. Then they give dozens of false alarms. I'm too busy to put up with another failing piece of complex equipment, so this device better not fail. At least not very often."

9. "You're all missing the point. The main thing is that this device has to coordinate elevator usage in peak hours, or else we'll have a revolution by our tenants. I know that we can do a better job of optimizing elevator use. I'm spending half my time answering complaints about the elevators."

10. "Yes, but that's all invisible to the passengers when it works right. I'm looking beyond that, to our image as the most modern office building in the downtown area. We need the artificial intelligence feature, so that people will be amazed that the elevator recognizes them and treats them personally, as if they were really special. That's what will attract the tenants."

11. "But if we start giving special treatment, we can't discriminate against anybody, or we'll have the government on our necks."

12. "And that includes the handicapped—we've got to be sure that everyone can use it, even if they can't see or hear."

13. "I think we're making this much too complicated. There are millions of elevators now in service, and they've been through just about everything. All we need to do is have this device communicate with other elevators in other buildings and learn from their experiences. How will it be done? I don't know. I'm no technician, but they can solve anything with computers these days."

18.2.3 Generalize the expectation list

Having obtained the list of specific expectations, we must now explore them to discover the full range of what is really expected. As we obtain this information, we revise and generalize the list, eventually producing a list that might include items such as these:

1. The device will provide upon request any and all information that is available, by computer modem, through North American telephone systems. This includes the latest weather information, stock quotations, sport scores, and news items.

2. The device, upon request, will provide individuals with simple health-related information such as blood pressure, pulse, and general state of stress and distress.

3. The device will warn individuals of potential health calamities.

4. The device will detect people thinking about committing a crime in the building and try to convince them that crime doesn't pay.

5. The device will diagnose elevator maladies and warn the appropriate people.

6. The device will quiet elevator passengers and controllable elevator sounds whenever the elevator passes quiet floors.

7. The device will soothe passengers experiencing anxiety in the elevator.

8. The device will be self-diagnosing and redundant enough to be completely fault-tolerant. It will replace any failed subsystem without loss of function and notify service personnel of the failed portion to be replaced.

9. The device will provide all current speed, direction, and floor stop command information to a central system that will moderate and coordinate optional commands on all building elevators in such a way as to make "optimal" use of each of the elevators. Optimal is to be determined by individual building managers, and may involve minimizing waiting time, energy consumed, complaints, or other such measures.

10. The device will determine the identity and travel patterns of all individuals who make frequent use of the elevator system, tell the individuals where the system "thinks" they want to go, and deliver them to that location by default. An override will be provided for passengers not wanting to go to the default location, and the system files will be updated. The device continually learns and monitors its own performance in adapting to its environment.

11. The device is capable of interacting effectively with all human beings independent of natural language, intelligence, age, intentions, or degree of physical impairment.

12.   All information provided by the device will be available to at least two of the three senses—sight, sound, and touch—depending on the individual receiving the information.

But what shall you do with item 13, which expects to have "this elevator communicate with other elevators"? You should be happy to get a few crackpot ideas, because the generalized expectation list will be a more useful instrument in the design process if it is created in a fantasy mode. This assures that nobody's expectations, no matter how wild, are omitted. We can indulge in this fantasy because we know that the expectation limitation process will put rational bounds on the flights of fancy.

18.2.4 Limit the expectations

Each item on the generalized expectation list can now be put into one of three broad categories:

P   *Possibility*, to be achieved now
D   *Deferred*, to be achieved in a later version
A   *Absolute* impossibility, to be dropped from consideration

Each different assignment of expectations to categories is likely to produce a substantially different design solution. Once again, the advantage of this process is in making explicit those decisions the designer would otherwise make implicitly. Making decisions explicit has to improve the match between expectation and realization of the design, or at least bring conflicts to the surface before too much time and money have been expended.

Sometimes, designers are reluctant to make decisions explicit because they fear the conflict that will result, but an essential part of design is the ability to communicate and then negotiate about disputed areas. By structuring the process of limiting expectations, we make it easier to deal with potential conflict in a rational way. For instance, suppose there is a dispute over the first item:

1.   The device will provide, upon request, any and all information that is available, by computer modem, through North American telephone systems. This includes, but is not limited to, the latest weather information, stock quotations, sport scores, and news items.

If the designer considers this to be too broad yet the clients consider it too limited according to their expectations, we would need to negotiate the content and rephrase the wording to make the perceived limitation quite explicit. We might finish with a restated expectation:

1.   The device will provide, upon request, any and all information that is available through AP, UPI, Reuters, Dow Jones, CNN, and ESPN wire services. These, along with local building sensors, are to be

the only sources of information external to the elevator itself in the initial implementation. To the extent possible, the design shall not prevent future extensions to other sources of information.

18.3 Limitations Need Not Be Limiting

Why is limiting expectations so important? Each different assignment to possible, deferred, or absolute produces many unique design solutions. If you don't make the assignment *consciously,* you will miss many design solutions, even many categories of design solution. If you do it *implicitly,* you *could* be consistent with your client's thinking—if you happen to be extremely lucky.

18.3.1 The wheelchair example

Even if you are lucky, you won't know for sure until much later in the design process—perhaps not until the final product is revealed. A wealthy lady, Mrs. Smythe, donated $14 million to a city in the Southwest for an art gallery. On the day of the opening, Mrs. Smythe was to cut the ribbon at the front door, but there was no way to reach it in her wheelchair!

The mayor publicly vilified the architects for not making wheelchair ramps an "achieve now" possibility (**P**). They defended themselves by saying that they assumed it was a "deferred feature" (**D**), to be remedied later—using funds from a second donation promised by Mrs. Smythe. Needless to say, her second donation was also deferred.

Part of the point of this story is that explicit limiting of expectations would have revealed the need for a wheelchair ramp. But that's only part of the story. Suppose it was really impossible to have a wheelchair ramp ready for the grand opening? Millionaires might be different from the rest of us, but most people would rather hear the bad news up front than feel cheated at the end. If you don't believe this, examine your own feelings.

Clients of the Elevator Information Device may not be happy that they have to settle for something less than all possible information on opening day. On the other hand, better—and more ethical—to let them experience the disappointment now, rather than surprise them when the system is delivered.

18.3.2 Keeping possibilities open

Not every expectation will result in a limitation. In rephrasing item 1, we were able to direct the designers explicitly to keep open the possibility of future extensions to the information system. Sometimes an entire item may be taken as a possibility, perhaps implemented now, or perhaps deferred. For instance, consider this:

2.   The device, upon request, will provide individuals with simple health-related information such as blood pressure, pulse, and general state of stress and distress.

Item 2 could be classified as a possibility and stand as written, which expands our thinking about the design problem to include this new dimension.

18.3.3 Include the source of the limitation

Of course, as the design proceeds, we may discover that item 2 can no longer be implemented because of legal, economic, or medical reasons. In that case, item 2 would be reclassified as an absolute limitation and rephrased to show the source of the limitation:

2.   *Because the cheapest medical diagnosis system now available costs more than $400,000,* the device will provide no health-related information of any kind to elevator passengers.

As we proceed through the entire list of expectations, some new possibilities will be added. At the same time, other possibilities will be transformed into limitations, so both designer and client will be clear and in agreement about just what the system won't do that it might otherwise be expected to do—*and why.*

Change is normal and natural, and this process of revision during the requirements process is no different from the revisions that take place later. But if we give *limitations without reasons,* the list becomes almost impervious to change. Suppose new and cheaper automated health information systems become available. If we hadn't documented the reason for limiting item 2, nobody would be aware that a new possibility was now open.

18.4 Helpful Hints and Variations

* In a way, working out the radar/computer problems was like translating the requirements into different foreign cultures and languages. In those stations where English was not the native language, certain documents did have to be translated. The translation process incidentally did a great job of revealing some hidden expectations.

  When you have a development project that will be carried out in or for two different cultures, you'll have plenty of problems arising from different cultural expectations. On the other hand, you can use the two cultures to your advantage by making sure that both are involved in the requirements process, in the same room at the same time, if at all possible. This may seem clumsy, but it's only doing work earlier and cheaper that would have to be done later at greater expense.

  Also, if you are going to have to translate requirements, don't wait until the requirements are finished before the translating begins. Have it done in parallel with the development of the requirements. You'll find that the exposure of differing expectations more than pays for the cost of translation.

- You can learn a lot about how requirements can create improper expectations by reading disclaimers, warranties, and special offers from Holiday Inn. (Just by following this suggestion, if you travel a lot, you'll find that this book will more than pay for itself. If not, you can get your money refunded at any *participating* bookstore.)

- A good exercise for revealing hidden expectations is the *Rule of Three:* If you can't think of three things that might cause your great idea to fail, all that means is that you haven't thought enough about it yet. For example, many designers get enthusiastic about the use of color coding in the Elevator Information Device. What effect will this have on blind people? What about color-blind people? people wearing coated lenses?

- Sometimes designers say, "If the clients knew about this limitation now, they'd kill this project. But after it's built, they'll like the other features so much that they won't mind that this one is left out. They probably won't even notice, so let's just keep it to ourselves and not make waves." Our advice is not to allow yourself to be tempted by this reasoning. It is the first step to losing track of who is the client and will lead to worse things than killing the project.

  What you can do, however, is pay attention to the process by which you *inform* the clients of this potentially disappointing limitation. You may be surprised. First of all, you may not understand just what it is the client expects. Or, the client may not care as much as you thought, or may be able to accept a reasonable alternative. Any of these outcomes would relieve the strain of secrecy and mistrust that otherwise would have infected the entire project. Or, the client may be a bit more clever than you think, and suggest a way of satisfying expectations that never occurred to you.

- Sometimes the limitation process can get bogged down in its own fantasies. Someone might say, "Yes, but suppose a blind person is standing on his hands, with his head in a bucket of water, and tries to operate the elevator controls with his bare feet. I don't think it will work." And then someone adds, "Yes, and suppose he's diabetic, and is waiting for his trained monkey to give him his insulin shot, but the system recognizes monkeys and won't allow them on the elevator."

  Of course, you don't want to suppress anybody's ideas, but your reluctance may give you a terrible time defining any limitations at all on an expectation. Sometimes, you can control this sort of fantasy gone wild by appealing to the *Doctrine of Reasonable Use,* which says, "We can't imagine all the bizarre things someone might do with this system, but we'll know what is reasonable when we see it. (And what do you think about having your head in a bucket of water?)"

- The Doctrine of Reasonable Use ultimately derives from the legal system. Questions of liability can be very real inhibitors to design, so if someone really thinks liability is at stake, you may simply have to invoke legal help. Quite often, merely asking, "Should we bring in an attorney on this issue?" will bring a sense of calm reality to the group—or else you may really have to pay for an attorney.

- Our colleague Eric Minch comments that the argument, "limitations without reasons...are impervious to change" can also be applied to functions, attributes, constraints, and all other descriptions of the product. At the same time, another colleague Ken de Lavigne observes that this explicit process of documenting limitations is radically different from what is common practice, which is to throw away all design materials after the code is written or, even worse, to keep them around without updating them. Ken then quotes a "British acquaintance" who remarked, "It's a pity that in many organisations (sic) the only cogent statements of business policy are to be found written in assembler language."

  As a practical matter, it may be unreasonable to expect that common practice will change so quickly, which is why we've emphasized the documentation of limitations as a start. Documenting why you're *not* doing something seems to be more palatable than documenting why you are doing something. Once people get a taste of the usefulness of documenting their reasons, the practice may spread to other descriptions of the product.

- We have found one way of documenting reasons that seems to be more acceptable and useful than anything else we've tried—videotaped interviews with designers and users. Many of the questions we've presented can be used, and people seem pleased enough to talk about their reasoning process. The videotapes can be dated and stored, and it turns out that people are very reluctant to throw away videotapes.

18.5 Summary

Why?
Designers need to be explicit with users about the limitations on their expectations because products and systems are much more readily accepted if the limitations are expressed honestly up front. Conversely, users feel cheated when they discover a limitation after the fact. Also, the process of developing and documenting limitations early in the design process helps to reveal important facts about the product or system.

When?
Raise the question of expectations whenever there is an indication of user dissatisfaction, as from the results of a user satisfaction survey. But even when there is

no obvious indication, conduct the expectations cycle as soon as the first round of requirements work seems to be drawing to a close.

How?

To raise and document expectations and limitations, follow this cycle:

1. Generate a list of specific expectations from representative users.

2. Work with the list to understand and generalize each expectation.

3. Negotiate to limit expectations to a reasonable level, leaving open possibilities for future modifications of the system, but definitely ruling out anything that can't reasonably be expected.

4. When setting a limit, be sure to document the source of the limitation, because today's limitation becomes tomorrow's opportunity.

Who?

Involve every category of user, in one way or another, so as to provide fuel for the expectation limitation process. When the limitations have been documented, place them in the hands of those users, to complete the cycle.

PART V   GREATLY IMPROVING THE ODDS OF SUCCESS

Barbara, Larry, and Todd, of BLT Design, are back in the chalk cave with Byron, Wilma, and John. They're more comfortable in the cave now, having spent many meetings there with these people, and they're beginning to regret that the Superchalk requirements phase is drawing to a close.

Barbara: We need to discuss how much work we have to do before we're done.

Byron: More work? Haven't we thought of just about everything?

Wilma: Not really. We haven't considered what will happen if someone steps on the chalk.

John: Or if someone tries to eat a piece.

Larry: Oh, come on, John, nobody would ever do *that!*

John: You've never taught high school.

Todd: He's right, Larry. We have to test for *everything*.

Larry: Test? We're still in the *requirements* phase.

Todd: Well, we have to *test* the requirements.

Byron: What for? I can see where we'll have to test the product, or even the design for the product, but why test the *requirements?*

Barbara: To see if we're really finished.

Todd: And I don't think we're finished.

Barbara: Why not?

Todd: Because I've just gone over the new draft requirements document again, and I've got a list of several hundred problems.

Byron: Let me see that!

(Byron grabs the list out of Todd's hand, glances at it, then shoves it back at Todd.)

Byron: I don't want your list of problems. I want the new draft requirements. Besides, your list looks like just a bunch of typographical errors—lots of commas and periods.

(Byron grabs the requirements document and starts to pore over it.)

Todd: Commas and periods are *very* important.

Byron: Well, it's more important that I see what you've drafted.

(While Byron is talking, Larry reaches over and gently removes the document from his hand.)

Larry: Trust us, Byron, my boy! Just trust us!

19   AMBIGUITY METRICS

Contrary to what Larry and Byron think, requirements do have to be tested. At least, they must be tested if the project is to have a fighting chance of success. In Part V, we'll develop specific ways in which requirements can be tested.

19.1 Measuring Ambiguity

The whole purpose of the requirements process is to reduce ambiguity in the development process, so the most basic test of any requirement is to measure its ambiguity. We've discussed requirements ambiguity, but the term "ambiguity" is itself ambiguous. There are many ways to reduce the ambiguity in "ambiguity," but the best would be to specify a precise way of measuring it. After all, there's little ambiguity in this requirement:

A.   *Draw a straight line on this page, 13.150 ± .025 centimeters in length, .500 ± .025 centimeters in width, using a Dixon Ticonderoga 1388 number 2 pencil. The line should be parallel to the top of the page, 2.000 ± .025 centimeters from the top, and touching the unbound edge.*

On the other hand, we've already seen how much ambiguity there can be in a requirement that says:

B.   *Design a transportation device.*

If "ambiguity" is a property of a requirement, then we'd like to be able to measure it in such a way that statement A has a small amount and statement B has a large amount.

19.1.1 Using the ambiguity poll

The ambiguity measure we will develop is based on something we already observed in the "Design a transportation device" example. We gave statement B to a thousand individuals, instructing them to work independently on a solution. As we saw, this problem statement lacks many essential ingredients, and we could have anticipated that each participant would resolve each one uniquely. Even if all solvers were trained to use the same design process, we'd be shocked to find that even two had produced exactly the same design.

Now suppose we gave statement A to a thousand individuals, working independently on a solution. The problem statement lacks a few essential ingredients, and since people are different, we would expect that some individuals would create idiosyncratic solutions. Generally speaking, though, we'd expect that there would be only a few variations, and that most solutions would be indistinguishable.

This mental experiment suggests that we can measure ambiguity as *the diversity of interpretation*. We have actually performed this experiment with large groups. For problem A, we measured the diversity by comparing the lines drawn. Everyone drew a single line, in pencil, and there were only minor variations in line length and placement on the page.

For the transportation problem (B), we gave each person one minute to develop a conceptual solution. At the end of that time, we asked each to give a single number as an estimate of the manufacturing cost of the proposed transportation device. Obviously, if they all had designed the same solution, their manufacturing costs would have varied somewhat, but not much. In fact, for about a thousand people, the cost estimates varied from $10 to $15 billion. This ratio of 1,500,000,000/1 indicates an extreme difference of opinion regarding the meaning of the problem, but isn't that precisely what we mean by ambiguity?

Such a ratio of largest to smallest estimate in a poll of informed individuals can be used as an *ambiguity metric*, a measurable entity for which we can obtain a precise value. Of course, a *precise* metric may not be extremely accurate, but it's far better than no measure at all. Indeed, it's a rather practical measure, one we have used often in real design situations.

19.1.2 Applying the polling method

A manufacturer of precision electrical components was studying the feasibility of developing an automated system for handling catalog information requests. The company's president had cost estimates ranging from "moderately inexpensive" to "moderately expensive," and didn't know whether to authorize the project. We asked each of 14 qualified individuals to write an independent estimate of the project cost. The estimated amounts ranged from $15,000 to $3 million. When the company president saw this 200/1 ratio, he halted the feasibility study to wait for a less equivocal requirement.

19.1.3 Polling on different bases

"Manufacturing cost" is only one possible basis for an ambiguity poll. Others that come quickly to mind are

- total design and development cost
- worker-years required for design and development
- number of unique parts in the solution
- minimum calendar time required to deliver the first product

These measures would apply to any project, but others would be more specific. For instance, in the transportation device problem, we might ask for estimates of

- *efficiency:* average energy consumption per hour of use
- *range:* how far it could transport
- *capacity:* maximum weight it could carry at one time

19.2 Using the Metric as a Test

One useful model of design says that *design is the process of removing ambiguity.* In terms of this model, design proceeds through a series of steps: creating an approximate design, testing for ambiguity, removing the ambiguity found, and retesting the new approximation. Eventually, the tests say that the approximation gets close enough, and design stops.

19.2.1 Measuring three kinds of ambiguity

In this view, everything that's done from recognition of a problem to final disposition can be considered design, including the unconscious, implicit design assumptions as well as the conscious, explicit design decisions. As an oversimplification, we can think of three major kinds of ambiguity that must be measured and removed, each associated with a major part of the total design activity.

1. *Problem-statement ambiguity,* as we know, is ambiguity in the problem statement, or requirements (see Figure 19-1).

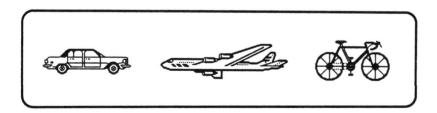

Figure 19-1. Problem-statement ambiguity is ambiguity in the *process that produces requirements* for the design. All three of these designs will meet the requirement to "design a transportation device."

2.   *Design-process ambiguity* is a measure of the variation in the process that will produce a design, in the sense of a picture of the solution (see Figure 19-2).

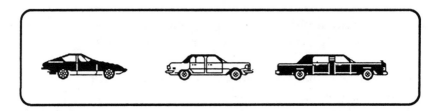

Figure 19-2. Design-process ambiguity is a measure of the variation in *processes that will produce a design* (that is, a *picture* of the solution). Three different processes to "design a car" could yield these three different cars.

3.   *Final-product ambiguity* is a measure of the variation in the process that will create the *physical solution,* that is, the product itself (see Figure 19-3).

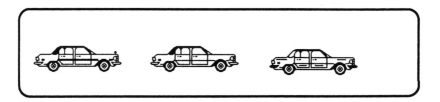

Figure 19-3. Final-product ambiguity is a measure of the variation in the *process that will create the physical solution.* For example, no two cars coming off an assembly line are ever identical, even though the specifications may be identical.

19.2.2 Interpreting the results

As a result of measuring these three types of ambiguity with our poll, you will know the amount of ambiguity remaining, which indicates the amount of design work remaining. You will also know something about which stage needs work to reduce the ambiguity. In the beginning, of course, this measure is largely a measure of the work necessary to hone the requirements—that is, the requirements ambiguity.

In our example, if we ask for estimates of "manufacturing cost of one device," the variations we get may stem from different understandings of this phrase. Some might interpret "manufacturing cost of one device" to imply a unit cost of volume manufacturing. In this case, design, development, and tooling costs may be insignificant because they are amortized over many devices. But other people might infer the production of only a single device, in which case, design, development, and tooling would represent a substantial part of the product cost.

This particular kind of ambiguity has little to do with the actual *product* of design—the motorcycle, bus, cruise ship, space shuttle, or skateboard that results. For instance, suppose two different estimators had a motorcycle in mind as the product emerging from the design process. If both interpreted "manufacturing cost of one device" as the unit cost in mass production, their actual production estimates might range from $200 to $4,000, an ambiguity of only 20/1 in the extreme. If both interpreted "one device" to mean the cost of development of a prototype, their estimates might range from $200,000 to $5 million, a 25/1 ambiguity. These ratios are far from the 10,000,000/1 that might arise from one person contemplating a space shuttle and another visualizing roller skates.

On the other hand, some of the ambiguity could stem from design-process ambiguity. Two people both might be thinking of cars, but one might be thinking of a car for a group of professional auto designers and the other for a soap-box derby competitor. How can we tell the difference? The simplest way is to ask the people polled.

What part of the problem statement led each of them to those inferences? This question will locate the ambiguous parts, so we know *what* to clarify and have a pretty good idea of *how* and *when* to do it.

19.2.3 Information from clustering

We can derive additional information from the ambiguity poll by noticing not just the total spread of estimates, but also how they bunch into clusters. In our survey of the transportation problem, there were three ranges that attracted large numbers of estimates: $40—$60, $300—$700, and $8,000—$15,000. Such clustering may indicate three common sets of simplifying assumptions.

When interviewed, the estimators revealed reasons for the extremes and clusters of assumptions shown in Figure 19-4. The vast majority of estimators assumed a *private* transportation system and never considered public transport such as planes, cable cars, or elevators. Most also assumed the device would have a large market, so that design and development costs would have minimal impact on their estimated unit cost. Neither of these inferences were explicit in the problem statement, yet most respondents felt comfortable with their own judgment.

19.2.4 Choosing the group to be polled

In this poll, the estimators were drawn from a population of professional systems designers with from three to thirty years of experience in the design and development of computer hardware and software. The great majority have spent their professional lives living and working in the United States, which might account for the assumption of private transportation, as well as for the skateboard, bicycle, and automobile clusters. It might also account for none of them imagining a camel.

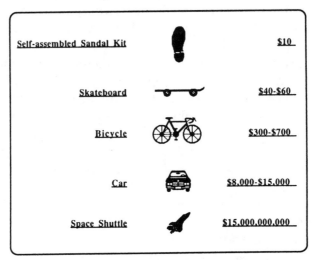

Self-assembled Sandal Kit		$10
Skateboard		$40-$60
Bicycle		$300-$700
Car		$8,000-$15,000
Space Shuttle		$15,000,000,000

Figure 19-4. The clusters of estimates may reveal assumptions about the potential solution.

These observations suggest that the most effective ambiguity poll is one in which there is a great diversity of respondents. The more differences among respondents, the more the poll will turn up different resolutions of ambiguity. Therefore, when measuring ambiguity, your first effort should be to find as diverse an audience as possible, even stopping people in the street and asking them the measurement questions.

19.3 Helpful Hints and Variations

- Our colleague Ken de Lavigne asks, "How about the expertise of the estimators?" With unskilled estimators, some of the variation will simply be due to errors in estimating, which is one reason to be sure to track down the source of the greatest and smallest estimates. Skilled estimators will avoid many mistakes, but in gaining the advantage of their skill, perhaps you lose variation—the more experienced they are, the more their experiences will tend to be the same. It's best to have some skilled estimators, but definitely not to have only skilled estimators.

- On the other hand, a lack of difference in the poll, even when the individuals are quite different, does not necessarily mean lack of ambiguity. For instance, it's possible to receive three independent estimates of $10,000 for the manufacturing cost of a transportation device, even when three different devices are being estimated. A high-tech, one-of-a-kind, anti-spill, collision-avoiding, floating-bearing skateboard might cost as much as an automobile, and certainly an Olympic racing bicycle falls in the same price category as a medium-sized car. This convergence of estimates explains why it's always a good idea to use several questions in the poll.

- One final word about polling as a method of exposing ambiguity: Polling will be effective only when the respondents' estimates are *independent*. If, for example, the company president's estimate is first given in public, then others are asked to present their estimates in public, you're likely to get a false convergence around the president's estimate. Any pressure, either overt or implied, will tend to reduce disparity in the estimates, and the less disparity, the less ambiguity revealed by the poll.

- Another colleague Eric Minch comments, "The paragraph on independence of estimates raises a large issue: the existence of biases and 'errors' in sampling and measurement is of interest to designers, too. Even if we could get perfect information on individual preferences, we'd still want to know what sorts of social dynamics are operative in consensus. This means we have to somehow tease out the influences at both levels from a single set of measurements. The social sciences have evolved a set of techniques for the design of experiments, from 'discreet' to 'clandestine,' which may be useful here."

  To this comment, we reply that we apply almost all of our measurement techniques in situations where we can watch people performing tasks. And, as that great social scientist Yogi Berra said, "You can observe a lot just by watching." Our own preference is to start simple before moving on to more sophisticated techniques. If simple's not good enough, however, you might want to look up some social science literature. Start with anthropology.

19.4 Summary

Why?
Use an ambiguity poll to estimate the ambiguity in a requirement, which measures *how much* work there is to do, and perhaps directs attention to *where* that work should be done.

When?
Whenever a piece of requirements work is said to be "finished," subject it to an ambiguity poll, to see if it really *is* finished.

How?
Conduct the poll as follows:

1. Gather a group of people to answer questions about the document whose ambiguity is to be measured.

2. Be sure that there is no pressure to conform, or no influence of any sort of one participant on another.

3. Propose a set of questions, each of which can be answered with a number, such as
    How fast?
    How big?
    How expensive?
    What capacity?

4. Estimate the ambiguity by comparing the highest and lowest answers.

5. Interview the high and low estimators to help locate the sources of the ambiguity.

Who?

The group used for estimating ambiguity should be as diverse as possible, at the very least including a sample from each population that will be affected by the eventual product.

20 TECHNICAL REVIEWS

There are two principal ways in which requirements information can be wrong: *inadequate* or *inaccurate*. Although most of our emphasis in this book is on adequacy—generating a sufficient *quantity* of significant ideas—that's only to balance our other work on accuracy—ensuring that each retained idea is of sufficient *quality*.

In this chapter, we'll give an overview of technical review meetings, the principal tools customers can use throughout the requirements process to test whether requirements contain all and only reliable information.*

All of what we have said about meetings in general in Chapters 8, 10, and 13 applies equally to technical review meetings, yet these meetings have a discipline all their own. The reason for the discipline is clear if you have ever attended a "review" like the one that follows.

20.1 A Walkover Example

(Jack, Zara, Martha, Sid, Ned, and Retha arrive at Room 1470B at 8:50 A.M. There is a carafe of coffee, a pot of hot water, and a huge pile of blueberry muffins. Everyone is in a good mood, and they attack the muffins. At 9:00, Jack steps to the front of the room, picks up one of several marker pens, and writes the agenda on the board: WALKTHROUGH AGENDA: WHAT'S WRONG WITH ZARA'S DO NOT DISTURB SIGN REQUIREMENTS?)

Jack: Okay, we're walking through what Zara did wrong on the sign requirements, so let's stay focused.

Zara: Don't I get to hear what I did right?

Jack: Sure, maybe if we have time at the end of the review. I mean, we all know it's a terrific piece of work, but you made a lot of mistakes.

Zara: Thanks.

---

*This chapter is but a short summary of what you need to know about technical reviews in order to run a successful requirements process. For a full guide, consult Daniel P. Freedman and Gerald M. Weinberg, *Handbook of Walkthroughs, Inspections, and Technical Reviews*, 3rd ed. (Chicago: Scott, Foresman and Co., 1983).

Martha: Well, personally, I wouldn't have made a sign at all, because signs get lost. I think the requirement should have had the sign built into the door.

Sid: Heavy.

Jack: What are you talking about?

Sid: Lost.

Martha: I think he means that a heavy sign won't get lost, or at least people won't carry it away with them by accident.

Retha: Handicapped people couldn't use it. Zara's requirement doesn't say anything about handicapped people, and that's essential.

Ned: Not really. Handicapped people are going to be noticed by the hotel staff, and they're not going to disturb them. Besides, even if the Do Not Disturb sign was out, and you hadn't seen this epileptic for two days, wouldn't you want to go into the room anyway and check it out?

Retha: Well, if the sign was built in, it could have a special time limit, kept by a digital countdown timer, after which it wouldn't be valid. Zara, your requirement doesn't say anything about time.

Zara: Well, I was trying to keep the expense down.

Sid: Parking cards.

Jack: What does that mean?

Martha: I think he means that the sign could have one of those little cardboard clocks, where you could set the hands.

Retha: That's still pretty expensive. And they'll get stolen.

(Retha stands up, goes to the board, and starts to write.)

Jack: What are you doing, Retha? I thought we agreed that I would be the one to write down all of Zara's faults.

Ned: Well, you're not writing anything.

Jack: Okay, okay. I'll write.

(Jack takes the pen from Retha and starts writing. Retha sits down and starts reading through Zara's requirements document.)

Sid: Language.

Jack: Can you hold that until I'm finished writing?

Ned: What are you writing, anyway?

Jack: I'm copying my list of Zara's faults. Then you can add yours to the list.

Ned: Well, how about giving someone else a chance? I've got a list, too. In fact, I've got a *long* list. I swear, Zara, your document is about the most ridiculous excuse for a requirement that I've seen in thirty years. Nothing personal, of course.

Zara: It sounds kind of personal.

Ned: Don't be stupid. You shouldn't get so attached to your work. I don't.

Martha: Lay off, Ned.

Ned: You don't need to defend Zara. She's a big girl.

(Zara spills her tea in Ned's lap, then stomps out of the room.)

Jack: Well, I guess the review is complete. I'm sure Zara will get over it soon. It's always a bit tough to have your work subject to a critical review, but I think it was well worth it. We've got this great list of faults to pass on to management. Retha, will you copy them down and have them typed?

Retha: Bleep!

Reader: As you study the following disciplines for successful technical reviewing, see how many violations you can find in this walkthrough of Zara's product that became a "walkover" of Zara.

20.2 The Role of Technical Reviews

The technical review is a testing tool, a tool for indicating progress of the requirements work. Technical reviews come in many variations, under many names, but all play the same role in managing the requirements process, as indicated in Figure 20-1.

20.2.1 Formal and informal reviews

To manage the requirements process, managers and even the producers themselves need reliable feedback on progress. That's the reason for having a review of each requirements document by *people who were not involved in producing it,* and this is known as the *formal* technical review. In an *informal* technical review, the producers of the document review their work critically *within their group.*

"Formal" and "informal," as we use the terms here, have nothing to do with how the meeting is conducted, but serve only to identify *where the output of the meeting goes:* results from the informal reviews to the producers only, and information from formal reviews to both producers and others—customers, project managers, or anyone else who needs reliable information on the progress of the requirements work. Inside an actual meeting, the two types of technical review may be identical in format. Indeed, an "informal" review could be conducted with a great deal more formality than a "formal" review of the same requirements document.

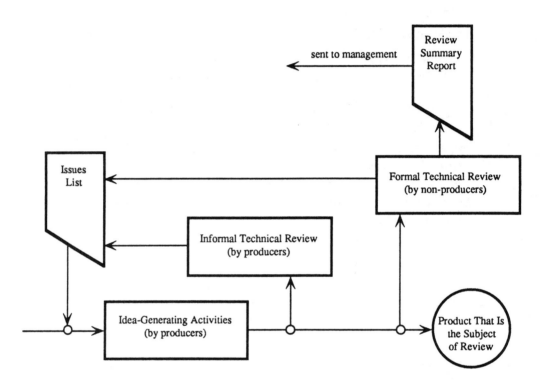

Figure 20-1. Technical reviews serve at least two functions in a project: feedback of issues to the producers to help improve the product, and feedback to management on the actual technical status of the project.

The other difference between the two reviews is the objectivity of the reviewers involved. While informal reviews are an excellent way to discover problems in a document, they are not always the best source of reliable information, because participants are less likely to discover their own mistakes and assumptions. In their brief meeting of the Do Not Disturb Project, for example, Jack and his coworkers managed to commit more than a dozen common reviewing errors, all the time seemingly unaware of what they were doing.

20.2.2 Technical reviews versus project reviews

Figure 20-2 illustrates the important distinction between a *technical* review and a *project* review, sometimes called a *management* review. Technical reviews include only experts on the subject matter and deal solely with technical issues. In reviewing a requirements document, these subject-matter experts are the customers or their representatives—the ones for whom the product is being designed. Their job is to provide information about the quality of the document. This information on quality goes to the document's producers, for them to make corrections.

A summary of the quality information from the technical review also goes to the project reviews. There, managers consider quality information along with information about costs, schedules, and resources in order to make informed judgments of what actions to take. The consequences of failing to control the quality of requirements are familiar enough—missed schedules, cost overruns, unmet requirements, inadequate performance, an error-prone product, and never-ending rework of the product. Only a proper combination of technical reviews and project reviews can prevent such outcomes.

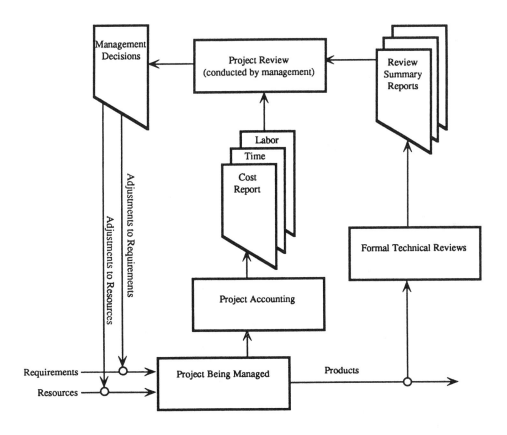

Figure 20-2. Technical reviews provide information to project reviews, which also consider information on nontechnical matters, such as resources consumed and available. The outcome of a technical requirements review would say nothing about such management issues as resources. The outcome of a project review, however, might be an adjustment of requirements to match resources, or of resources to match requirements.

20.3 Review Reports

Whatever goes on inside a requirements review, its major project control function is to provide to interested parties such as customers, users, and management a reliable answer to this fundamental question:

Does this requirement do the job it's supposed to do?

That answer is conveyed in the review reports.

20.3.1 Need for review reports

Once any piece of the requirements document has been reviewed and found acceptable, it becomes part of the ongoing requirements process. Therefore, within a very small risk factor, the document must be   .

1.   complete

2.   correct

3.   dependable as a base for related work

4.   measurable for purposes of tracking progress

Without technical reviews, there are no *reliable* methods for measuring the progress of the requirements work. *Sometimes* the producers themselves give a reliable report, but *sometimes* is not good enough. No matter how good their intentions, producers are simply not in a position to give consistently reliable reports on their own products.

For small, simple projects, with well-intentioned, competent producers, there is a finite chance of success without reliable measures of progress. As projects grow larger and more complex, however, working without reviews is like exploring the jungle without a compass. The chance of at least one self-report being overly optimistic is no longer a chance but a certainty. And if the requirements are mistakenly thought to be correct, the rest of the project will never emerge from the jungle alive.

The reports issued by a formal review are like a compass for management. They serve as a formal commitment by competent and unbiased people that a piece of the requirements work is complete, accurate, and dependable. By themselves, the review reports do not guarantee that a project will not end in failure or crisis, any more than a compass guarantees you won't get lost in the jungle.

One of our clients, after a very poorly done requirements review, remarked, "Around here, we call a review like that a 'tates.' " When we looked puzzled, he explained that Colonel Stoopnagle on the old radio show "It Pays to Be Ignorant," once invented a compass that pointed in random directions, which he name a "tates." "And that's because," he said, *"he who has a tates is lost."*

Trying to manage a requirements process with "tates" is sure to get you lost, but well-done review reports are not in themselves sufficient to make your process succeed. To keep the process on track, the project's management must use the information in the review reports to make informed management decisions about what direction to take.

20.3.2 Creating the issues list

As a review meeting progresses, it raises and records *issues* with the requirements document, which become an *issues list*. This list is essential because it tells the producers why their work was not totally acceptable, ideally in sufficient detail with sufficient tact to enable them to remedy the situation. Figure 20-3 is an example of an issues list from a requirements review.

```
Issues List from Review RQ-17.32.3-1   27 October

1.   Standards.  The meaning of "user interface
standards" does not seem to correspond with the
common use of that term throughout the rest of the
project.

2.   Incorrect References.  The reference numbers to
other documents seem to be incorrect.  A list of
possible corrections is attached.

3.   Incomplete Explanation.  Some reviewers felt that
the explanation of why alternative C was dropped was
not sufficient, because it gave no indication of how
the decision would change if critical cost factors
for outside suppliers were changed over time.

4.   Missing Diagrams.  Three diagrams were given only
in data flow notation, whereas the standard calls for
Warnier-Orr diagrams in addition.  The other six
diagrams were in both notations, though with slight
errors in the data flows (see annotated copies
attached).



_____          Lynn Wabash, Review Leader



_____          Merle Chaumiere, Recorder
```

Figure 20-3. An example of a requirements review issues list. No special form is needed, but special care is required to write accurate and effective issues.

The issues list serves primarily as a communication tool from one technical group to another. As it is not intended for nontechnical readers, it need not be "translated" for their eyes. It need not be fancy, as long as it is clear.

A common subversion of the review process is an attempt to make the issues list into a solutions list. The job of the review committee is only to raise issues; it is the job of the producing unit to resolve them. A review committee is generally no better at resolving issues than a producing unit is at raising them. Even when the producers are reviewing their own work, attempting to produce solutions—idea generation—at the same time they're raising issues—idea reduction—is a one-way ticket to an unfocused meeting.

20.3.3 Technical review summary report

The one report that is always generated by a *formal* review is the *technical review summary report*. This report, intended for management and perhaps customers, carries the committee's assessment of the work, which is derived from a weighted opinion of the seriousness of the items on the issues list. Thus it is the fundamental link between the review process and the project management system. For effective project management, review summary reports must identify three items (see Figure 20-4):

1.   What was reviewed?

2.   Who did the reviewing?

3.   What was their conclusion?

An informal review obviously does not need a review summary report, though in practice many informal reviews go through the discipline of making an appraisal of the work unit as acceptable or not acceptable, for what reasons, and to what degree. This final appraisal is essential because no one can know, just by counting items on the issues list, how significant the issues are.

20.4 Principal Types of Review

There are many different kinds of reviews, with varying formats and purposes. In this section, we'll describe some of the most common variations.

Technical Review Summary Report

Review Number_____ Starting Time_____
Date   _____ Ending Time_____
Requirement Identification RQ-_____
Produced By_____
Brief Description_____
_____

Materials Used in the Review *(check here ☐ if supplementary list)*
    Identification   **Description**
_____   _____
_____   _____
_____   _____
_____   _____
_____   _____

    Participant   **Signature**
*1.*_____   _____
*2.*_____   _____
*3.*_____   _____
*4.*_____   _____
*5.*_____   _____
*6.*_____   _____
*7.*_____   _____

Appraisal of the Work Unit:
    Accepted *(no further review)*   **Not Accepted** *(new review required)*
    ☐ *as is*   ☐ *major revisions required*
    ☐ *with minor revisions*   ☐ *rebuild required*
                                          ☐ *review not completed (explain)*
    Supplemental Materials Produced   **Description/Identification**
    ☐ *Issues List*   _____
    ☐ *Related Issues List*   _____
    ☐ *Other*   _____

Figure 20-4. A sample requirements review summary report form. The form should fit on one page, with supplementary pages added if necessary to document exceptional situations, so that the status of a piece of work can be taken in at a glance.

20.4.1 *Vanilla reviews*

A requirements review can be conducted without any particular meeting discipline decided in advance, simply by adjusting the course of the meeting to the demands of the product under review, like adding toppings to vanilla ice cream to create a do-it-yourself sundae. This type of "vanilla" review can be very effective because of its adaptability, but for this reason, it requires a more skilled facilitator than some of the more structured varieties.

Over time, the project culture will tend to evolve special disciplines that emphasize certain aspects of reviewing at the expense of others, and also develop more skilled review leaders. Unfortunately, these developments usually come too late to help the requirements process. That's why for requirements work, it's a good idea to have a pre-planned review discipline and/or to retain experienced facilitators to lead the reviews.

20.4.2 Inspections

Many of the best-known review disciplines are attempts to cover a great quantity of material in the review. For example, the "inspection" approach tries to gain such efficiency by focusing on a narrow, sharply defined, set of questions. In some cases, an inspection consists of running through a checklist of faults, one after another, over the entire document.* Obviously, one danger of such an approach is from faults that do not appear on the checklist. That's why effective inspection systems generally evolve methods for augmenting checklists as experience grows.

20.4.3 Walkthroughs

Another way to cover more material is by having the product "walked through" by someone who is very familiar with it—even someone who specially prepares with a more or less formal presentation. Walking through the product, a lot of detail can be skipped—which can be good if you are just trying to verify an overall approach or bad if your object is to find errors of detail.

In some cases, the walkthrough is very close to a lecture about the product, which suggests another reason for varying the review approach. In some cases, rapid education of large numbers of people may suggest some variation of the walkthrough. But, as always, it's a good idea to make clear from the outset that the main purpose of the meeting is education, not idea reduction.

20.4.4 Round robin reviews

In a walkthrough, the process is driven by the product being reviewed. In an inspection, the list of points to be inspected determines the sequence. In vanilla reviews, the order is determined by the flow of the meeting as it unfolds. In contrast to these types, *round robin reviews* emphasize a cyclical sharing by the various

---

*See Daniel P. Freedman and Gerald M. Weinberg, *Handbook of Walkthroughs, Inspections, and Technical Reviews*, 3rd ed. (Chicago: Scott, Foresman and Co., 1983).

participants, with each taking an equal and similar portion of the entire reviewing task.

Round robin reviews are especially useful in a situation often found in requirements work, when the participants are at roughly the same level of knowledge—a level that may be as high as can be obtained, but not as high as is needed. This might be the case when a large number of different users need to review the requirements, each from a different perspective. The round robin format ensures that nobody will shrink from participation through lack of confidence, while at the same time ensuring a more detailed look at the requirements, part by part.

20.5 Real Versus Ideal Reviews

If you observe actual reviews, you will never find one following all the "rules" of its particular type. That's because each type of review, like any other tool, has its own advantages and disadvantages, so that any real review contains some sort of mixture of all types. Let's take the walkthrough as an example to illustrate why real reviews always vary from the ideal.

Because the presenter has prepared in advance, a walkthrough can move through a great deal of material at high speed. Moreover, since walkthrough reviewers are far more passive than those in other varieties, large numbers of people can attend the meeting and become familiar with the walked-through material. This larger audience can also bring many diverse viewpoints to bear on the presented material. If the audience is alert, and represents a broad cross section of skills and viewpoints, the walkthrough can give strong statistical assurance that no major oversight lies concealed in the product.

The walkthrough does not make demands on the participants for advance preparation. In much requirements reviewing, there often need to be large numbers of participants, in order to include the views of all different users of the product. These users may come from diverse organizations not under the same operational control. In such situations, it may prove impossible to get everyone prepared for a review, so the walkthrough may be the only reasonable way to ensure that all those present have actually looked at the requirements.

So much for the *advantages* of a walkthrough. As with all the review varieties, the *problems* of the walkthrough spring rather directly from these unique advantages. Advance preparation by all participants is not required, so everyone may have a different depth of understanding. Those close to the work may be bored and not pay attention. Those who are seeing the requirements for the first time may not be able to keep up with the pace of presentation. In either case, the ability to raise penetrating issues is impaired.

In order to retain their attention, the review leader may, at intervals, ask participants in turn for comments, thus introducing a round robin element to the meeting. In order to keep up with complex material, individuals may bring their own private checklists, thus adding an element of inspections to the meeting. Rather than try to suppress these "deviations," use them as information about how the review tool needs to be sharpened.

20.6 Helpful Hints and Variations

- Everything we have said in the previous chapters about meetings as tools naturally applies to requirements reviews as well. For instance, the requirements review should be prepared to produce a *report of related issues,* to help keep the meeting on track.

- The reports from technical reviews, when analyzed together, can provide an important historical overview of the project so far. They can reveal patterns among issues over time, which can guide management in changing which tools are used, how tools are used, and what kind of training is given to project participants. To make historical analysis possible, create a file of all summary reports and issues lists, and try to ensure that they are done in comparable formats. Don't, however, let the formatting needs of historical analysis put a dead hand on the live process of doing the reviews.

- Nobody likes to be criticized, and it is far too easy to believe that a technical review is criticizing you as a person, rather than your product. A good set of agreed-upon rules helps get through this delicate time, as does a talented corps of review leaders and a set of experienced reviewers. If people are reluctant to have their products reviewed, or even attend reviews, that's an indication it's time to look at the rules you use, the leaders you use, and the reviewers you invite.

20.7 Summary

Why?
Requirements reviews are needed because no one can produce error-free requirements on a consistent basis, and because the producers are the least likely people to see the errors in their own products.

When?
Use requirements reviews whenever you need to determine whether a requirements document does indeed do the job it was intended to do.

How?

1. There are many review disciplines from which to choose. Try them all, and adapt them to your particular needs.

2. Allow time for learning, and have patience with your awkwardness.

3.   Develop a feedback system so that what you learn in early reviews is used to improve future reviews. The easiest way to do this is to develop a trained corps of review leaders.

Who?
Choose review leaders for their interpersonal skills and their special training in the various review disciplines.

Select review participants on the basis of their ability to detect issues in the product, as well as their ability to handle themselves appropriately in what can be a most delicate situation. Customers and users are the primary candidates for participation, but you may need some skill and tact to decide which customers and users are to be invited. This is particularly true of certain *combinations* of users, and you may wish to conduct separate reviews of certain requirements documents just to keep the meetings from becoming battlefields for political factions.

21 MEASURING SATISFACTION

Perhaps ninety percent of product development efforts fail. About thirty percent fail to produce anything at all, but most of the failures don't have that problem. They do produce a product, but people don't like it. They may not use it at all, or if they do, they may grumble endlessly.

21.1 Creating a User Satisfaction Test

The easiest way to avoid user dissatisfaction is to measure user satisfaction along the way, as the design takes form. Designers can also be considered users, which is another reason to employ a *user satisfaction test* as a communication vehicle between designers, as well as among clients, end users, and designers.

21.1.1 Test attributes

The test we use was inspired by an approach suggested by Osgood, et al.,* but sufficiently modified so that Osgood need not take any blame for what we've done. It was designed to have the following attributes:

1. It allows user satisfaction to be measured frequently, so that changes during the design process can be detected and evaluated.

2. It provides the designer with a periodic indication of divergence of opinion about current design approaches. The designer can identify strong differences of opinion among informed individuals, differences that warn of unrecognized ambiguity or political problems.

3. It enables the designer to pinpoint specific areas of dissatisfaction so that specific remedies can be considered.

4. It enables the designer to identify dissatisfaction within specific user constituencies.

*See Charles E. Osgood, et al., *The Measurement of Meaning* (Urbana, Il.: University of Illinois Press, 1957).

238

5. It enables the designer to determine statistical reliability of the sample, which is important because we usually don't have the luxury of surveying the entire population of users.

6. Because it is enticing to the user to complete, it generally musters a high percentage of respondents.

7. It provides the designer with a clear understanding of just how the completed design is to be evaluated.

8. Even if the results are never summarized, the test is useful to those who fill it out. In fact, even if nobody fills it out, the process of creating the test itself will provide useful information to the designer.

9. The test is cheap, easy to use, nonintrusive, and educational to those who administer it and to those who fill it out.

21.1.2 A custom test for each project

The test is the same in form, but different in content for each project. To create the user satisfaction test for a particular project, we ask the clients to select a limited subset (six to ten) of the original set of attributes by which the final product will be evaluated. For example, the user might select the following for the design of a Do Not Disturb system:

    multicultural
    inexpensive
    easy to use
    offering few options
    highly reliable
    easy to service
    having easily understood information

From this subset, we create a *set of bipolar adjective pairs,* one adjective representing the most favorable condition and one representing the least favorable. These are organized on a seven-point scale, with the least favorable on the left and the most favorable on the right. In our example, the test would look like Figure 21-1. Notice that the large area of white space, conspicuously entitled, "Comments?" is often the most significant part.

When the design of the test has been drafted, show it to the clients and ask, "If you fill this out monthly (or whatever interval), will it enable you to express what you like and don't like?" If they answer negatively, then find out what attributes would enable them to express themselves, and revise the test.

When the clients have eventually answered this question affirmatively, the designer knows that these seven attributes are to play a crucial role in the evaluation of the final product. The designer will keep these attributes in the foreground, rather than lapsing into implicit assumptions about how the design will be evaluated.

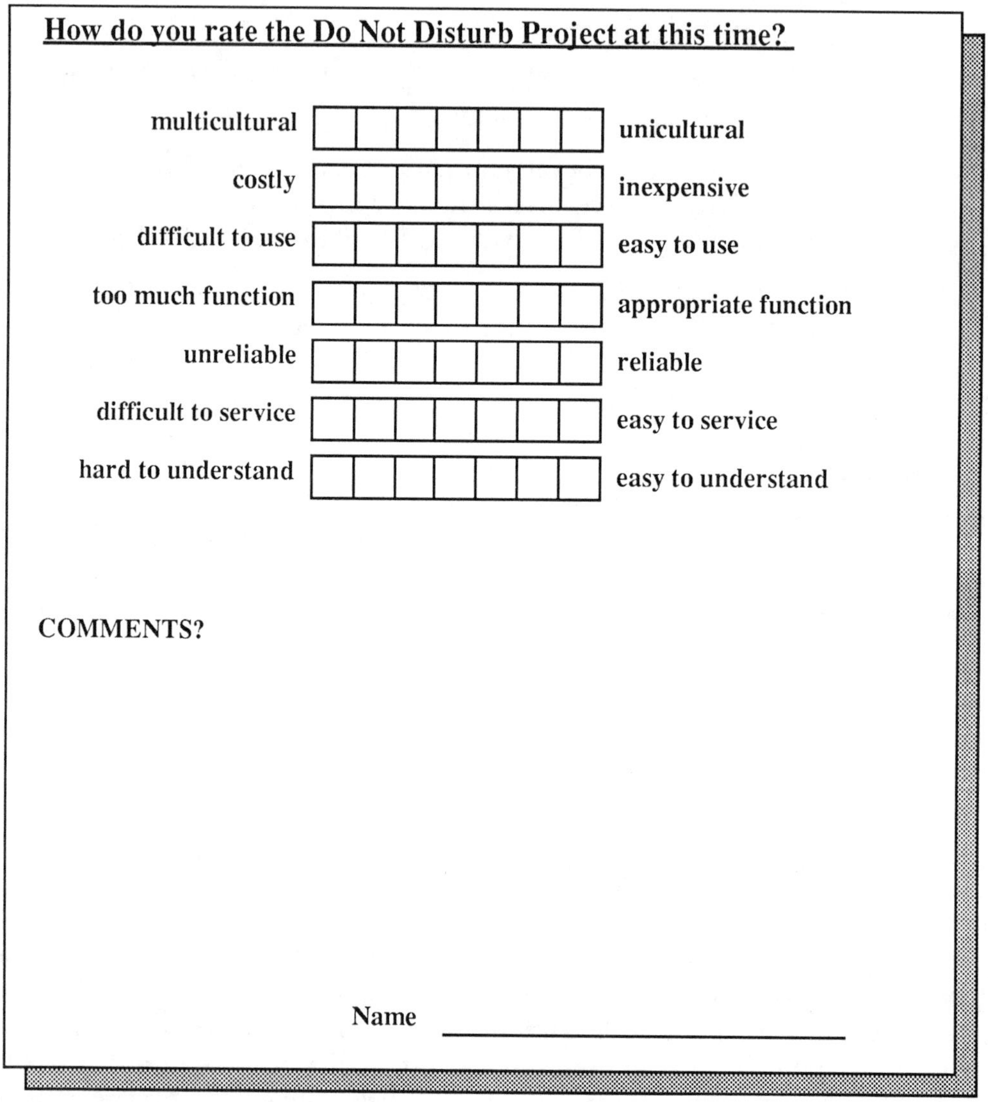

How do you rate the Do Not Disturb Project at this time?

multicultural	▢▢▢▢▢▢▢	unicultural
costly	▢▢▢▢▢▢▢	inexpensive
difficult to use	▢▢▢▢▢▢▢	easy to use
too much function	▢▢▢▢▢▢▢	appropriate function
unreliable	▢▢▢▢▢▢▢	reliable
difficult to service	▢▢▢▢▢▢▢	easy to service
hard to understand	▢▢▢▢▢▢▢	easy to understand

COMMENTS?

Name _____

Figure 21-1. A user satisfaction test for the Do Not Disturb Project.

21.2 Using the Test

The user satisfaction test can be applied throughout the requirements process and on into the rest of the project.

21.2.1 Benefits

In designing the test itself, we use it as a tool to find out what are the important and less important attributes. We would have this benefit even if the test were never administered to anyone.

Filling out the test regularly, however, helps to keep the client involved and focused on what is really important. It also helps the design team determine how consistent their parts of their approach are with the client's wishes.

We can also use the test as a constraint on the design process. The requirements document itself could state that all attributes must have an average score of, say, at least 5.0 when administered to a specified sample of users. This, however, is not the principal use of the satisfaction test.

The greatest importance of the test lies not in any absolute significance of the numbers, but in its ability to spot changes in satisfaction. Thus, the two most important things to look for in the responses are *changes* and *additional comments*, as discussed below.

21.2.2 Plotting shifts and trends

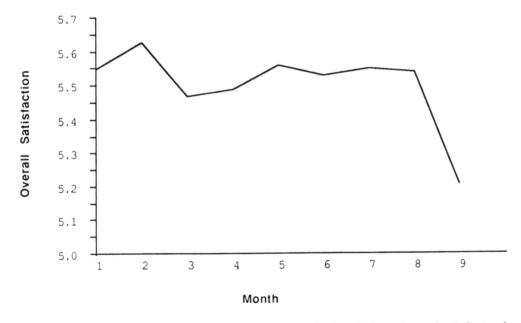

Figure 21-2. Regular plotting can reveal shifts in satisfaction levels, which can be used to indicate where further information is needed.

You may wish to regularly print, and post, plots of average satisfaction levels, as indicated in Figure 21-2. A shift—up or down—in the average satisfaction rating of any attribute indicates that *something* is happening. This shift is sufficient cause for the designers to follow up with an interview. Sometimes the shift is the result of no more than a general mood of depression or elation, but that in itself is good to reveal. At other times, though, the shift will be an early indication of something that might cause big trouble if allowed to continue unnoticed.

For example, Figure 21-2 shows what happened late in the requirements phase of a project to create a CAD/CAM (computer-aided design/computer-aided manufacturing) system. The average overall ranking dropped from around 5.5 to 5.2. Considering the number of people in the sample, this was a significant drop. It was traced to very severe drops in satisfaction by three engineers. When interviewed, they revealed that in a meeting to work out limitations, the one feature that was absolutely essential to their work was deferred to a later modification.

According to them, the person running the limitations meeting did not understand why their work was different from the other engineers, and did not listen to them when they tried to explain. The specific deferred item was easily raised to a current requirement, but the response didn't end there. Further inquiry revealed that this particular facilitator had acted the same way in other meetings, though none of the people who had been disturbed happened to be in the satisfaction sample. The facilitator accepted an invitation to get further training in facilitation skills.

21.2.3 Interpreting the comments

Listening in a meeting is much like reading any comments on the user satisfaction sheet. When people take the time to write something free-form, you can be sure it's important. But it's not always the thing they say that counts, but what's behind it. That's why puzzling comments must be tracked down.

For instance, one respondent said, simply, "This form stinks!" When interviewed, the respondent said he didn't like the form because he wasn't comfortable writing down some very delicate issues, issues he was more than happy to discuss in person. Translated, "This form stinks!" meant, "I don't know how to use this form to convey information I think is very important. Please come and see me about it!" That's what most "bizarre" comments usually mean.

21.2.4 Feelings are facts

Some years ago, we gave a traveling seminar entitled, "Measuring and Increasing User Satisfaction." Hundreds of people attended, and many were stunningly enthusiastic. We were surprised, however, to find that about thirty percent of the attendees got extremely angry when we presented the user satisfaction test.

Inasmuch as this was a case of user dissatisfaction with user satisfaction, we decided to apply our own methods. (Of course, we used a test to measure satisfaction with the seminar, which is how we discovered the problem in the first place. See Figure 21-3.) It turned out that most of the dissatisfied attendees equated "user satisfaction measurement" with "performance measurement"—the study of automatically measurable characteristics of a computer system, such as the distribution of response times for various interactions.

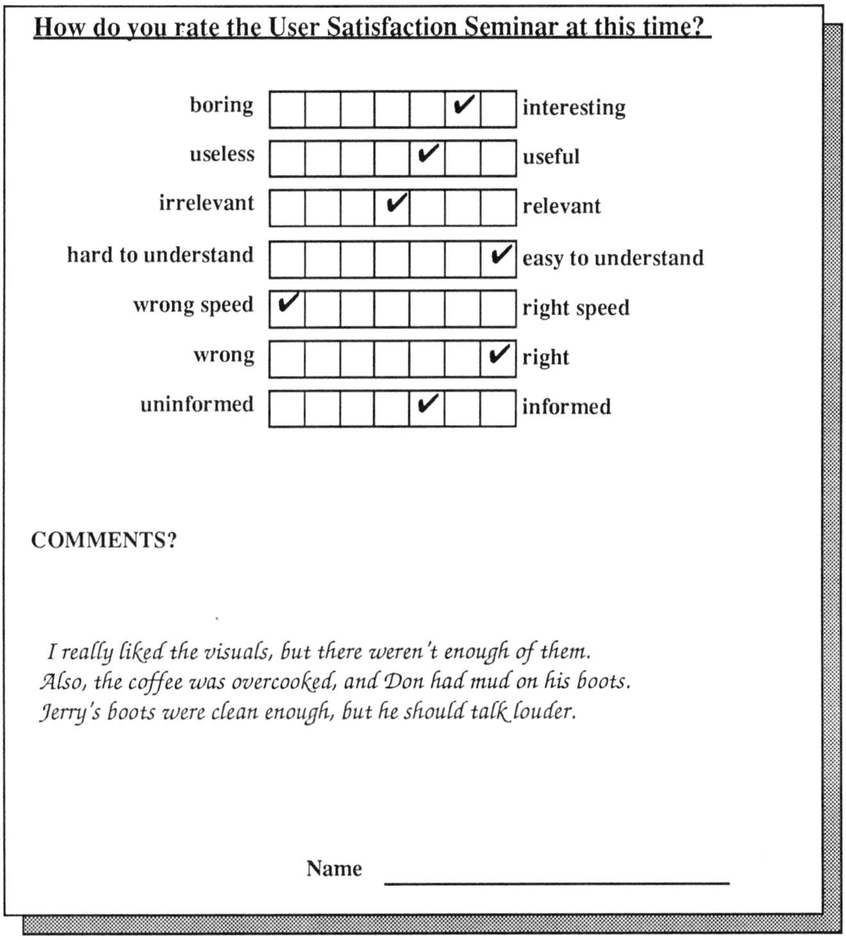

How do you rate the User Satisfaction Seminar at this time?

boring	☐☐☐☐☑☐	interesting
useless	☐☐☐☑☐☐	useful
irrelevant	☐☐☑☐☐☐	relevant
hard to understand	☐☐☐☐☑☐	easy to understand
wrong speed	☑☐☐☐☐☐	right speed
wrong	☐☐☐☐☑☐	right
uninformed	☐☐☐☑☐☐	informed

COMMENTS?

I really liked the visuals, but there weren't enough of them.
Also, the coffee was overcooked, and Don had mud on his boots.
Jerry's boots were clean enough, but he should talk louder.

Name _____

Figure 21-3. A user satisfaction test for seminar evaluation.

As one performance measurer said, "What do I care whether they say they are satisfied with response time? They signed off on a document that agreed they would be satisfied with ninety percent of the responses in less than one second, and that's what my measurements show. They have no rational cause for complaint."

It may be that they have no rational cause for complaint. It may even be that they have no legal leg to stand on. But they are the customers, and their feelings are facts, just as much as the percentage of one-second response times is a fact. And, when it comes to designing new things, subjective reactions are the most important facts of all, because they warn us that our design assumptions are not turning out as planned.

If you don't think feelings are facts, important facts, ask yourself the following questions: Why do customers often risk legal action and monetary penalties by breaking contracts for products that don't satisfy them? And why do they switch suppliers for future products?

In the above case, one of the performance measurer's users was standing next to him when he made his statement. "That's right," she said, with a trace of bitterness, "we did agree that ninety percent of the responses under one second would be satisfactory. But we assumed that the other ten percent would be more or less like in the old system—no longer than thirty seconds in any case. With your new system, sometimes I have to sit for ten minutes waiting for a response. . . ."

"That's not true," he interrupted. "We've never measured a response time of ten minutes."

"Since I have no warning when it's going to happen, I've never measured it. It may not actually be ten minutes, but it seems that way. Whatever it is, it's not satisfactory, and if you don't fix it, we're going to go back to the old system."

"But it's not my fault. I built what the requirements called for."

"It *is* your fault, for not helping us write a meaningful requirement," she insisted, terminating the conversation by turning her back on him.

Don't ever allow yourself to dismiss a response you get on a user satisfaction test as "just feelings." If it weren't for feelings, nobody would ask for a new product to begin with.

21.3 Other Uses of the Test

The user satisfaction test has many benefits, independent of its use as a satisfaction measuring device.

21.3.1 A communication vehicle

Because it has what psychologists call "face validity," the test is especially useful as a communication vehicle. "Face validity" means there are no psychological arcana needed to interpret its meaning. Like any good communication device, it says what it means and it means what it says.

The average response measures major shifts in satisfaction, but face validity means that even the response of a single user can be used as a source of information. Figure 21-4 plots the highest and the lowest satisfaction ratings, over time, creating an "envelope" within which all responses lie. It would be easy to ignore these extremes and just go for the average, but the maximum information will be gleaned from those who rated the extremes.

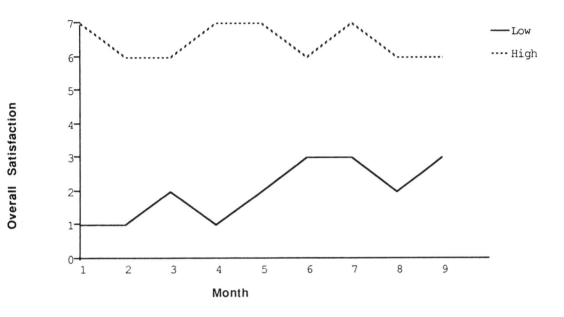

Figure 21-4. Plotting the envelope of high and low responses indicates where we might go for maximum information.

You can also use maximum *shifts* to identify users with a lot of information. If a user changes "easy to use" from 6 to 3, for example, that means the user no longer thinks the product is easy to use. If another user changes the reliability scale from 0 to 6, it means that most doubts about the system's reliability have dissipated. In either case, follow up with the user to discover the reasons for the change.

21.3.2 Continued use throughout the project

Another benefit of the test is that it can be continued in use, unchanged, after the product is delivered. The results then become a measure of how well people are learning to use the product, and how well it is being serviced. They also provide a starting point for initiating follow-on projects.

21.3.3 Use by designers

The user satisfaction test provides an additional service for teams of designers: It tells them how their performance is going to be measured. Without such a focusing device, each designer is likely to emphasize an idiosyncratic perception of the desired solution. Since millions of assumptions are possible, the final integrated product would undoubtedly lack a certain consistency with the real customer requirements.

In the design phase, the test can also be used to measure satisfaction among the design team itself. If the designers in a large project use the test regularly, it will measure convergence among designers on what the design problem really is. Divergence provides an early indication of areas of misunderstanding, and an automatic excuse to sit down and discuss them.

21.4 Other Tests

There are, of course, many other ways to measure user satisfaction as a development project proceeds. The existence of a regular survey should never be used as an excuse for not supplementing it with other, commonsense methods. For instance, you can always take the opportunity simply to ask users, "How do you feel about the system right now?" They'll probably respond ninety percent of the time with an automatic, "Just fine," but the other ten percent may lead to a gold mine of information.

21.4.1 Prototypes as satisfaction tests

If the product is being built incrementally, of course, the best measure is the product itself as far as it has been built. That is, you simply observe how people actually use it, avoid it, or modify it, and record this information to show what satisfies and what doesn't. This is the entire foundation of design using prototypes: You not only build a prototype, but you see what people actually do with it.

Of course, you may not realize until later that you were building a prototype. A classic case in computing was the design and implementation of the very first FORTRAN compiler. Although there are few people around today who remember this, about half of the effort in that first implementation went into the FREQUENCY statement. Few FORTRAN programmers today have ever heard of it, but yes, it truly did consume half of the effort.

The FREQUENCY statement was to be required after an IF statement if the compiler was to use the IBM 704's two arithmetic registers and three index registers efficiently. In those days, code efficiency was a paramount issue for potential users, but as people began actually using FORTRAN, they discovered that the greatest inefficiency was not in the compiled code, but in the compiler itself. A typical program would be compiled ten or twenty times, during debugging and run only once. Quite often, the total compile time was a hundred times the final running time of the program, and users quickly discovered the reason why. Most of the compiler's time was spent analyzing FREQUENCY statements for optimization.

One of the first "enhancements" of that first FORTRAN system was to use a "sense switch" on the console to allow the user to turn off frequency optimization. In those days, all you had to do was look at the setting of sense switch #1 to survey the user satisfaction with FORTRAN's FREQUENCY optimization feature. That's why you never heard of the FREQUENCY statement.

If the FORTRAN developers had been more aware of the use of a prototype to measure satisfaction, they might have first built the compiler without the FREQUENCY feature, then gotten some customer reaction before investing so much in a feature that was never used. That's easy to say in hindsight; the trick is to have foresight on *your* next development project.

21.5 Helpful Hints and Variations

• When there are two or more rather distinct client groups, create a user satisfaction test for each group. Each test will have its own adjectives, and the different groups may find it informative to see the others' tests. The designer, too, will find this exercise useful for raising ambiguities and potential conflicts.

• During the life of a long project, you can normally expect a small, steady decline in measured user satisfaction. People get bored, or interested in other things, so that nothing quite matches the excitement of a new project. So don't be unduly concerned by such a gradual decline, but, on the other hand, consider whether you need to do something—not too severe—to revive flagging interest.

• Be sure to choose your sample large enough so it won't be overly affected by one individual. If one individual has a real gripe, you'll see it in the comments, or just as a sharp change in the individual rankings.

• Be sure to maintain the size of your sample throughout the life of the project. If people leave, they must be replaced to keep the samples reasonably comparable.

• If you have trouble handling bizarre comments, you may want to put an additional check-box on your user satisfaction test:

     ☐ I have more to say. Please see me or call.

• To create a statistically significant survey, seek the advice of a competent statistician, but first you may want to ask if your survey needs to be statistically significant. Most of the time, it will be more important just to keep the information flowing, and follow up on any changes with interviews. Don't get bogged down in mathematics. You're after usable information, not publishable theories.

• Ken de Lavigne warns that if you do use a prototype as a satisfaction measuring device, "Don't sell it!" He's speaking from experience in the software industry, but in many industries such as aerospace, the first prototype is so expensive, it also has to be the production model. In those cases where the prototype is not quite so expensive—when, say, designing furniture—you can sell it as a designer original. The important thing is not to forget the information function of the prototype in the quest for maximum bucks.

21.6 Summary

Why?
The easiest and only reliable way to ensure that users will be satisfied with the ultimate design is to measure their satisfaction as the design takes form. Without regular measurement, it's too easy to engender the response of the American who went to Paris and ordered *tête de veau* in a three-star restaurant. When he saw the veal head's beady little eyes staring at him from the platter, all he could say was "What's that?" He never did manage to eat it.

When?
Create the user satisfaction test as early in the project as possible, soon after the attributes are listed for the first time. Administer it on a regular basis, with the interval depending on the project's lifetime. On a one-year project, monthly surveys would seem about right.

How?
Follow this cycle:

1. Create a user satisfaction test for your own project. Use our form or some other that suits the culture of your organization. And above all, be sure to get user involvement in the process.

2. Administer the test regularly, as promised.

3. Tabulate each category and the overall satisfaction rating, then look for shifts. Follow up on the shifts to find out what's behind them.

4. Pay particular attention to any comments, especially if they express strong feelings. Never forget that feelings are facts, the most important facts you have about the people for whom you're creating this system in the first place.

5. Whether or not you use a user satisfaction test, don't forget to use all the information about user satisfaction from other sources, such as reactions to prototypes, or simply off-hand comments.

Who?
All users should at least be represented in the sample. Be sure to replace people who leave the purview of the project. Use the test with the design team, too, but keep the tabulations separate. If the measures differ widely, call a meeting of the designers and users to find out why.

22   TEST CASES

Classifying expectations is one way to test requirements. The user satisfaction test is another way—the ultimate way, in fact. But these are not the only ways and given our imperfections we need as many ways as possible. Surprisingly, to some people, one of the most effective ways of testing requirements is with test cases very much like those for testing a completed system.

22.1 Black Box Testing

22.1.1 External versus internal knowledge

In systems theory, the concept of the *black box* (or *opaque box*) is used to study a system whose inside, for one reason or another, is not available for inspection.*
Because the box is "black," all that's known is what goes in and what comes out: the *input* and the *output*. By trying various inputs and examining the resulting outputs, you can learn *what* the box does, but nothing about *how* this transformation is implemented. There might be electronic circuits hidden inside, or a machine, or a squirrel in a cage.

We can use the black box concept during requirements definition because the design solution is, at this stage, a truly black box. What could be more opaque than a box that does not yet exist? Through our explorations, we already know a great deal about the box, such as its name, the answers to some context-free questions, who will use it, and what behavior those users consider important. All of this knowledge is *external* knowledge of our design solution, not knowledge of its inner workings. Therefore, we are still dealing with a black box.

Design itself is the process of converting a black box to a *white* (or *transparent*) *box*—one in which we can clearly see all the details of *how*. Exploring the system from a stimulus-response point of view is a way of describing the black box (the requirements): "When we do that, this must happen." (Figure 22-1)

---

*For more information on black boxes, see Gerald M. Weinberg, *Rethinking Systems Analysis & Design* (New York: Dorset House Publishing, 1988).

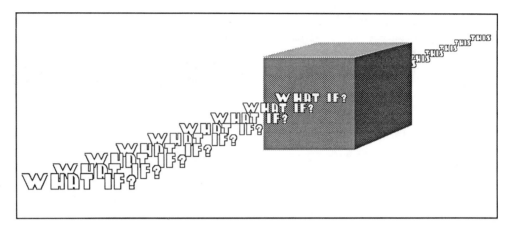

Figure 22-1. **The black box is an imaginary sealed box containing the final product. We put in "What if" questions and get out "This will happen" answers.**

For example, consider the way an attorney draws up a will. After you say you want all your money to go to your nephew's children, the attorney asks, "What if your nephew doesn't have any children?" You never thought of that possibility, so you think awhile, then tell the attorney if that happens, all the money should go to a fertility clinic. After a few iterations, the attorney has all your wishes covered—even some you didn't know you had—and draws up (designs) your will.

A will is like a black box test to the extent that it refers to *events in the future* that may not even happen, like your nephew's having children. Requirements are just that kind of future black box. To test them, we must imagine things that *might* happen, even if they have never happened before. In fact, because no product exactly like this product has ever been built, the events could not have happened before, so we have to reason by similarity to other experiences.

22.1.2 Constructing black box test cases

To learn more about the Elevator Information Device black box, we'll construct some *test input cases* as if the system already existed, then see if we can describe how the output should look. With the client, we construct "What if" questions such as the following:

1.   What if a person selects the correct command when entering the elevator?

2.   What if a person selects the *incorrect* command when entering the elevator?

3.   What if an unauthorized person attempts to invoke a secured command such as obtaining personal information, directing the eleva-

tor to a secured floor, or modifying the information presentation formats?

4. What if an act of violence takes place in the elevator?

5. What if there is a power failure in the building?

6. What if a passenger has a medical emergency?

7. What if it is unsafe or extremely unpleasant on any of the selected floors? (Examples: bad weather on observation floors, riots of aggressive shoppers, a hold-up in progress in a bank, long waiting lines, or floors still under construction.)

8. What if a person on the elevator wants a taxi or other services that would traditionally be provided by a concierge?

9. What if the passenger is severely impaired in one of the three senses —sight, sound, or touch?

10. What if someone gets on the elevator who doesn't know where to go?

11. What if the prospective passenger is befuddled by the complexity of the system's interactive mechanisms?

12. What if an act of vandalism occurs on the elevator?

22.1.3 Testing the test cases

Notice how one test case seems to lead to another. Even if we never attempt to answer the test cases, just developing them provides a solid test of our requirements work up to this point. But we will attempt to answer them, because answering them helps to improve our requirements process. For one thing, answering provides a *test of the test cases themselves.* The question is, "Have we constructed a good and sufficient set of test cases?"* As we shall see, trying to answer the black box questions will also help to answer *this* question.

Initially, though, we want to test whether our list of test cases has *coverage.* Are all the functions of the system "touched" by at least one test? We can rather easily check that the major functions are covered by simply listing which functions each test is designed to check.

But coverage means more than just touching a function. We must also touch that function *throughout its range of behavior.* That means the tests must consider

1. normal use of the function

2. abnormal but reasonable use of the function

3. abnormal and unreasonable use of the function

---

*For more information on checking test cases, see Daniel P. Freedman and Gerald M. Weinberg, *Handbook of Walkthroughs, Inspections, and Technical Reviews*, 3rd ed. (Chicago: Scott, Foresman and Co., 1983).

For instance, writing a geometry lesson on a blackboard is clearly normal use for Superchalk. Drawing on clothing is not normal, but is quite reasonable to expect. Eating Superchalk may be unreasonable, but the design will have to deal with this issue in some way, in order to prevent lawsuits. No single failure of requirements work leads to more lawsuits than the confident declaration, "No sane person would do *that!*"

Figure 22-2. "No sane person would do that!"

"Unreasonable" refers to what sane people would *not* do. The best way to be sure that you've tested for the unreasonable is to listen for arguments that you shouldn't include the test.

22.2 Applying the Test Cases

After a fairly complete set of test cases has been constructed, we attempt to describe the output for each case. We can ask the users to construct these answers, or construct them ourselves and then present them to the users for verification and elaboration. We can conduct a face-to-face session, or survey the users for written responses.

22.2.1 Examples

The answers should be of the form, "If X happens, the product should do Y." Here are some examples for the Elevator Information Device:

1. If a person selects the correct command when entering the elevator, the system will immediately respond by indicating that the command was understood, and how it was understood by the system.

2. If a person selects the *incorrect* command when entering the elevator, the system will immediately respond by indicating that some error was made, if it is at all possible to discern that it was an error. The elevator information system should be able to discern as much information as a human operator who has experience with the building. For instance, we would not expect the system to know that a passenger intended to go to the fourteenth floor if the passenger actually specified the fifteenth floor, unless there was some further information to indicate that this passenger would not want to go to the fifteenth floor. We do not expect all errors to be discernible by the system, so the system should allow the passenger to correct a self-detected error. For instance, if the passenger says "fourteen" and then realizes that she wanted to say "fifteen," she should not be forced to stop at fourteen.

3. If an unauthorized person attempts to invoke a restricted command such as obtaining personal information, directing the elevator to a secured floor, or modifying the information presentation formats, the system will record the event, time, date, and, if possible, the identity of the person attempting the command. The command will not be carried out, and the person who issued it will be notified of the illegal command and the recording process.

4. If an act of violence takes place in the elevator, the system will detect the act, immediately alert building security, begin recording the event on videotape, and attempt to intervene by warning the individuals involved. The elevator will move directly to the security floor, and the doors will open only when requested from the outside.

5. If there is a power failure in the building, the information system will contain its own power backup system that will provide elevator occupants with light, ventilation, and information on the source of the problem and what they should do. The system will inform security of how many people are on which elevators, and where.

6. If a passenger has a medical emergency, the system will ask for medical information, provide useful information to occupants, notify medical emergency personnel to stand by, and rush the elevator to the designated emergency floor.

7. If it is unsafe or extremely unpleasant on any of the selected floors, the system will notify the selecting passenger of the conditions, provide information for alternate sources of service, and, except in cases

of actual danger, allow the selector to re-select the same floor. In cases of danger, the elevator may require identification and a security check before allowing the re-selection, so as to allow police, fire, emergency, and other authorized workers to reach the hazardous floor.

8. If a person on the elevator wants a taxi or other services that would traditionally be provided by a concierge, the system will notify the concierge so that the service will be available at the appropriate time.

9. If the passenger is severely impaired in one of the three senses—sight, sound, or touch—the individual will be able to fully communicate with the system and command the elevator with equal effectiveness using either of the other two modalities.

10. If someone gets on the elevator who doesn't know where to go, the system will allow such a passenger to easily inquire of the system, and the system will provide all necessary information through a dialogue.

11. If the prospective passenger is befuddled by the complexity of the system's interactive mechanisms, there will be an obvious (obvious even to the most naive or sensory-impaired person) HELP command, which will provide soothing stimuli and assistance. If the prospective passenger is still anxious and unable to command the elevator in a desired manner, the elevator will transport the passenger to the nearest designated HELP floor where a human being is available for assistance.

12. If an act of vandalism occurs on the elevator, the system will sound an alarm in the elevator and in the security office. The elevator will move directly to the security floor, and the doors will open only when requested from the outside.

22.2.2 Iterating tests and answers

Do some of these answers sound unreasonable to you? If so, you are in the same position as a designer who hears unreasonable answers from a client, or a client who hears unreasonable suggestions from a designer. That's exactly why we recommend using the black box exploration: not so much to provide answers as to raise new questions.

As the test cases are answered, new test cases spring to mind, and we repeat the process using these test cases. For instance, the answer to the question "What if the passenger is severely impaired in one of the three senses—sight, sound, or touch?" led to the requirement:

9.1.   The individual who is severely impaired in one of the three senses—sight, sound, or touch—will be able to fully communicate with the system and command the elevator with equal effectiveness using either of the other two modalities.

This requirement, in turn, led to two more questions:

9.1.1. What if the passenger is severely impaired in two or more of the three senses—sight, sound, or touch?

9.1.2. What if the passenger does not read?

The answers given were

9.1.1. If the passenger is severely impaired in two or more of the three senses—sight, sound, or touch—the system will not necessarily interact effectively with such an individual, but hazardous conditions will be avoided.

9.1.2. The system must provide nonreading passengers with only the basic control and information elements provided in today's standard elevator.

The second reply led to yet another question:

9.1.2.1. What if the passenger is not an English speaker?

This was answered in the following way:

9.1.2.1. The system must provide command and information facilities with equal ease to English, French, German, Spanish, Chinese (Mandarin), and Japanese speakers. For passengers speaking none of these languages, the system must provide only the basic control and information elements provided in today's standard elevator using suitable international symbols.

After a few rounds of black box exploration, both designer and client have a better intuition for what is desired and what is reasonable to expect.

22.2.3 Clearly specifying ambiguity

Sometimes we've been accused of destroying the creative possibilities of design, because we overconstrain the designers by being so specific about just what the system should do in all conceivable cases. In effect, each answer to a black box question does become a constraint on the designers, but it's never our intention to overconstrain. And, it won't happen if you pay attention to the idea of *clearly specifying ambiguity,* or don't care conditions. For instance, notice the answer given to question 9.1.1:

9.1.1. If the passenger is severely impaired in two or more of the three senses—sight, sound, or touch—the system will not necessarily interact effectively with such an individual, but hazardous conditions will be avoided.

Careless writing might have said, "the system will not interact effectively...," instead of saying "not necessarily." In other words, the answer clearly specifies that effective interaction is not required, but the customer doesn't care—and won't object—if it does. Thus, the requirement doesn't restrict the creative designer from finding a way of helping the multiply impaired person.

Once the list of answers has been stabilized, take the group through the list one more time, ideally with the assistance of at least one designer and a professional writer. For each answer, ask, "Does this overconstrain? Does it inadvertently specify more than we really mean?" For example, we re-examined our answer to question 12:

12.   If an act of vandalism occurs on the elevator, the system will sound an alarm in the elevator and in the security office. The elevator will move directly to the security floor, and the doors will open only when requested from the outside.

The designer raised a question about the word "sound." Did the alarm have to be audible, or was it restricted to being only audible? After some discussion, the answer was revised to read:

12.   If an act of vandalism occurs on the elevator, the system will provide an effective alarm in the elevator and in the security office. (The design of this alarm should be tested and approved by the security committee.) The elevator will move directly to the security floor, and the doors will open only when requested from the outside.

22.3 Documenting the Test Cases

The results of the black box test cases can become the basis for systems and product testing later on. Knowing this, however, we may be tempted to put off developing these black box cases until some later time, perhaps when an official systems testing group has been assembled. This is a mistake. You must not postpone the black box work, for it begins to lose its effectiveness if you have already started thinking about the mechanisms for accomplishing the objectives. In that case, the systems test may fall into the trap of testing that the mechanisms do things right (design and implementation tests), but not that they do the right things (requirements tests).

Instead, you must do the black box testing while still at the stage of defining the problem, not giving the solution. You are attempting to understand specifically what the client wants to do without ever making commitments as to how it is to be accomplished.

Developing these black box tests and replies early in the process and documenting them as requirements for acceptance tests is the surest way to avoid both the designer saying that the requirements have changed and the client saying that the requirements have not been met.

22.4 Helpful Hints and Variations

- There is an entire body of folk literature all over the world consisting of "three wishes" stories. These are all stories of what happens when your wishes are fulfilled, and none of them are pretty. Perhaps this is the ultimate test of the problem definition. Have the group imagine that all wishes in the requirements have come true—all constraints have held, all objectives have been optimized, and all expectations have been met. Then what happens?

- Writing a user manual is an example of the black box approach to requirements. The user manual doesn't show internals, nor should it show them, because by "user" in this context, we mean people whose interaction with the system implies never "going inside." The people who do go inside will also be users, of course, but they will have their own "internals" manuals. Even so, these internals manuals may not allow them inside *everything*. There may indeed be many levels of user, each with an idea of what is black and what is white.

  In many organizations, development of the user manual is deferred until after the system is at least partially built, which loses at least two important opportunities. First, you lose the use of the user manual as an additional requirements test. Second, the user manual is much more easily infected with *how* things are done, because that's what the writers are looking at when they develop the manual. If the manual is developed when the design is still inside the black box, though, there is no danger of such an infection.

- Some people resist black box testing of requirements with the argument that this opens the danger of "studying for the test." They fear that the designers will build a product or system that does no more than these tests specify, rather than one that will be wise enough to cover cases nobody thought of. Well, if the designer knows other important questions and their proper answers, why not put them on the list? There they will be subject to public scrutiny and also be available for constructing acceptance tests.

  We'd certainly like our product or system to be wise, but where does such wisdom come from? If the product that meets all requirements isn't wise enough to satisfy us, then we weren't wise enough when we developed our requirements. What makes us think we'll get magically wiser later on? Sure, we learn things as we proceed with the design, but when that happens, we must be sure to backtrack and incorporate them into the requirements, lest we expose ourselves to side effects and unmet expectations.

- Some people like to use black box techniques very early in the requirements process, to generate ideas about product functions. There's nothing wrong with this, and there's nothing to prevent the process from being resumed once you've gotten further in the requirements process.

- It's not cheating to look at acceptance tests for other products to get ideas for black box tests of your requirements. In fact, it's never cheating to do anything just because it gives you a fresh supply of new ideas.

- Of course, some things *are* cheating—like breaking into your competitors' offices late at night and rummaging through their files. But you're not cheating because you get new ideas, you're cheating because you're cheating within some larger environment. You shouldn't break the laws of God, State, or Nature to get your ideas, but you can *fantasize* breaking those laws and see what new ideas come up. For instance, what do you *think* you might find at midnight in a competitor's office? What could happen if someone used your product for immoral purposes? Best to think about it now, before it happens!

22.5 Summary

Why?
Construct and answer black box tests primarily to test the completeness, accuracy, clarity, and conciseness of any requirements developed so far. You may also use them as the basis for later constructing actual tests of the product or system.

When?
To be most effective, black box test construction must be done before you start designing solutions. In some cases, it can be done very early, as a way of getting the requirements process moving.

How?
The process of black box testing of requirements follows this cycle:

1. Construct a series of test cases by imagining that the product has been built and asking "What if" questions.

2. Answer the test cases, and discuss the answers with all interested parties in an effort to obtain agreement.

3. Attempting to reach agreement on answers will generally raise other "What if" questions. Add these to the list and answer them, repeating the process until the list stabilizes.

4. Once the list is more or less stable, scrutinize it with a team that includes designers and professional writers. This team is specifically searching for overconstrained replies, then revising them to give the greatest possible design freedom—but no more.

5.   When the revisions have been made, document the cases so they become the starting basis for constructing systems and acceptance tests.

Who?

Anyone who will have a say in using, accepting, or rejecting the product should be involved in test case generation, and especially in constructing answers. In addition, designers and professional writers are very helpful at rewriting the answers to avoid overconstraint.

23   Studying Existing Products

Almost any good book on systems analysis will teach the reader how to use information from existing products as a requirements starting point.* Therefore, instead of duplicating work that is well done elsewhere, we want to describe the use of existing products *as an ending point*. In this chapter, we'll discuss how to use the existing product to test the adequacy of the requirements job.

23.1 Use of the Existing Product as the Norm

Thousands of years ago, Aristotle said, "It is not once, nor twice, but times without number that the same idea appears in the world." Now, as then, very few "new" products are entirely new. And, given the nature of people, almost any product you design will eventually be compared with something that already exists. Moreover, it will be compared in terms of the metrics developed for the existing products, not on its own merits.

Alexander Graham Bell may have invented the first telephone, but he had to suffer endless unfavorable comparisons with the telegraph. Today it may seem obvious that the telephone is a superior device in almost every way, but it wasn't obvious then. The standard of comparison for telegraph systems was "correct words received per minute," over a single line by skilled operators. Indeed, skilled telegraph operators could send and receive correct words at least as fast as they could be heard over a primitive telephone and written down.

Fortunately for Bell, he was aware that his invention would be held up to unfair comparison with the telegraph, and much of his energy was spent finding other measures for comparison. His colleague, Watson, was a tenor of some talent, so Bell used to have him sing over the phone to impress would-be investors. Singing couldn't be done at all over the telegraph, and since Watson sang familiar songs, his listeners didn't really notice that they couldn't hear all the words very well.

---

*See, for example, Stephen M. McMenamin and John F. Palmer, *Essential Systems Analysis* (Englewood Cliffs, N.J.: Prentice-Hall, 1984).

Figure 23-1. New products are compared to old in terms of the metrics developed for the existing products, not on their own merits.

Whether you start or end the requirements process by looking at existing products, then, the new product will have to survive a comparison with those products. Since the existing products will be used as norms, regardless of whether or not they are appropriate norms, you need to use them as *tests* of requirements.

23.2 Interviewing

Just because you have interviewed certain users during requirements work doesn't mean you can't use them again as a test when a draft requirement is finished. This is especially true when you speak to representatives of users, because you merely have to choose a different set of representatives for the test. This approach also tests how good the representation was. But even if you interview the same people before and after, there are many useful tests you can perform, as we describe below.

23.2.1 What's missing in the new product?

In practice, the first thing people notice about a new product is whether it is missing anything that they have come to expect in the products it is perceived as replacing. A few days after our colleague Stiles Roberts sent us a critique of an early draft of this book, he called to say, "I don't remember a requirement in the Elevator Information Device to indicate whether the elevator was about to go up or down." When reading the manuscript, he hadn't noticed the omission (did you?), but after riding in an ordinary elevator, the lack of an indication suddenly became obvious.

It's the same psychology that forces up the rent every time we move to a new apartment. The new apartment *must* have everything the old one had, or we will sorely miss it, or so we think initially. As it's unlikely that it will have all and *only* what the previous apartment had, we will also pay for the new features. These new features, in turn, become requirements in our next apartment.

Thus, the first requirements test using existing products is to *ask about missing functions*. Not all the missing functions will be oversights; there may be functions we don't want the new product to supply. But we will at least want to be prepared to explain *why* those functions are missing, and *how the user is to do without them*. If we aren't prepared, the users will be sure to complain, and may reject the new product without ever appreciating its superiority.

23.2.2 Why is it missing?

Of course, we may actually find some oversights, in which case we will want to add them to the requirements documents. If there are many oversights, we will want to ask ourselves, "What deficiency in our requirements process led us to overlook so much? And what *other* problems may have arisen from that deficiency?"

Some years ago, a publisher commissioned a new information system to keep track of royalty payments to authors. When the requirements team believed their work was complete, they called upon Jerry to lead a technical review of the documents. He invited the six clerks who handled the present system to review the document, specifically comparing it to what they now did. The first thing they noticed was not the wonderful new features, but the lack of any procedure for correcting errors.

The systems analysts who had prepared the requirements argued that the new system wouldn't make errors, and thus did not need any special error-correcting procedures. The clerks pointed out that only a few of the errors arose from their work, but instead originated in faulty reporting from the bookstores, as well as from misinterpretation of contract provisions.

Jerry then asked, "Why did the analysts leave out correction procedures?" The answer was that they had only examined the interface of this system with one bookstore—the one the publisher ran themselves—and hadn't talked seriously to any authors. The requirements process was reopened, and an error-correction function was added. Perhaps even more important, as a result of the analysts' talking to a few authors, the entire royalty report was redesigned, much to everyone's satisfaction.

In this example, the test against the existing product revealed a flaw in the user-inclusion strategy. But even when the user inclusion has been more thoroughly done, it's easy to miss many functions. For one thing, interviewees can't always think of everything when put on the spot. For another, in the early stages of a project, the interviewers don't fully appreciate the significance of every answer they hear. It's always safe to assume that something will be missed, and to schedule a test at the end.

23.2.3 What's missing in the old product?

Users are ordinarily very aware of certain annoying shortcomings of a product, and this information can be obtained through fairly direct questions, such as,

* *What doesn't work in the present product?*
* *What groups of people have what troubles with the present product?*

On the other hand, people who use a product can adapt to almost anything, and may no longer notice what's missing. Questions such as the following will tend to tease out this "forgotten" information:

* *How much time do people spend on various parts of the present product?*
* *How much money do people spend on various parts of the present product?*

23.2.4 What's missing in the old requirements?

Of course, there's an even easier way to find out what's missing in the old product: *Look at the requirements for the old product and compare them to the product that was actually produced.* When you make this comparison, you will learn something about the process by which requirements are translated into products. There is a reason for each missing requirement, and a reason for each function that's in the product that wasn't in the requirements. If you can track down these reasons, you may learn a great deal about what you need in the present requirements.

Near the end of the requirements work of an electronic mail system, for example, we discovered a copy of the requirements for the existing system. One "required" item that was conspicuously missing was a broadcast function that enabled a user to transmit a message to every one of the system's users. The new draft requirements called for a similar function because some users had requested it. We asked one of the old-timers why it hadn't been implemented in the old system.

"Oh, but it *was* implemented," he told us. "After the system had been operating for a few weeks, someone broadcast a technical question about the system itself."

"That sounds like the kind of usage we were trying to foster," we said.

"Well, about twenty people had answers to the query, so each of them broadcast an answer. Then there were another twenty or so who didn't agree with each of those answers, and then there were those who didn't agree with their disagreements. By the time the thing died down, we figured that there were more than a hundred thousand messages transmitted because of that one broadcast."

"So you removed the broadcast function?"

"Well, let's say we made it *harder*. You can still broadcast, but you'd have to make an explicit address list of every single person, which discourages swamping the system."

This little interview saved us from making the same costly mistake a second time, but we wouldn't have been able to ask the right questions if we hadn't run across a copy of the old requirements. It would be nice if companies routinely saved the requirements documentation for every product still in use, but sometimes you have to be a bit more creative about getting this kind of information. Here are some sources we've used successfully at least once in the past:

1. "Pack rats" are just likely to have an unofficial copy of the old requirements. Ask people "Who saves everything?" to help locate the pack rats. Jerry once found a twenty-three-year-old requirement this way.

2. Old news releases or articles, sometimes in-house organs, often have lists of features, at least. They can usually be found in complete sets in company libraries. If there is a company librarian, you have a powerful ally in your search.

3. Old advertisements and marketing literature for products are another source. The early releases are usually based on requirements, rather than on the actual product, and so reveal discrepancies between the two.

4. You can interview old-timers, but they're usually not too reliable unless they're prompted. At least try asking, "Can you remember anything that was promised but was never delivered? How about something that was in the new system that was later taken out?"

23.3 Substituting Features for Functions

Looking at existing systems is like opening a can of worms. In the worst case, you will tend to add *features* to match the existing product, rather than the *functions* you really want. This tendency is especially pronounced when product marketing people get into the requirements act with a *competitive analysis*—a list of all possible features for all possible competing products. By implication, they want all of these features to become requirements in the new product.

Figure 23-2. It should be easy to sell a Rolls Royce with a Volkswagen's price and gas mileage, but can you build it at a profitable cost?

If your sales force can say that your product has every feature of the competitor's product, then you can make the sale. This idea is sound enough. It should be duck soup to sell a Rolls Royce clone with the price and gas mileage of a Volkswagen (see Figure 23-2).

But can you build such a car and make a profit? Every time you add a feature, it tends to add *constraints* to the other attributes. Ten thousand blades will provide great competition for the Swiss Army knife—if you can keep the weight under ten pounds and the cost under a hundred dollars.

So how do you keep your product from becoming a Swiss Army product? Going through all the heuristics can help, but the most effective step is to create functions at the right level of description. Instead of trying to compete with everyone on a feature-by-feature comparison, try to identify the functions that will be needed to compete, then you can develop the attributes of those functions in the usual way.

For instance, suppose you are building a new car, and the competitive analysis says that the other cars in this class all have gold monograms on the steering wheel. When the marketing people try to make "gold monogram on steering wheel" into a requirement, ask, "What *function* does it serve?"

Most commonly, you'll get the reply, "It doesn't serve *any* function. It's just there to match the competitor's."

"In that case," you say, "the function is 'match the competitor's.' And just what is it about the competitor's we're trying to match?"

"Their pizzazz! Their sales appeal!"

"Great. So we can say that the function we want is 'match the competitor's sales appeal.' "

"Sure, but how are we going to do that?"

"Possibly with a gold monogram on the steering wheel, but possibly not. Perhaps the designers will give us a platinum monogram on the dashboard. Or perhaps they'll produce a combination of price, performance, and pizzazz that will do the trick. That's up to the designers to decide. Our job is to tell them our requirements."

23.4 Helpful Hints and Variations

- One particularly helpful test is to ask people to reminisce about the early days of the previous product, if anyone was present and can remember that far back. To the extent that the culture hasn't changed, these memories will give you hints about features that are needed to get the product accepted for use in each environment.

- A related approach is to try to discover what entire *product* is missing. Whenever we're working with an organization for the first time, we ask about *other* projects that never finished products, or finished products that had to be withdrawn from use. Tracking these down always yields a barrelful of information.

- Upgrading present products tends to produce Swiss Army knife-type systems. To serve the existing population of users, you must retain all the old features as well as add new features to keep them happy. But to win new users, you must keep the product simple and cheap, but these attributes will tend to be destroyed by a Swiss Army of features. At some point, your best choice may be to split the product into two related products—a simple, unburdened one to attract novices, and a sophisticated one for jaded old-timers. Knowing when to split a requirements process into two is a great art and a possibility you should always keep in mind when the feature list starts to get unwieldy.

23.5 Summary

Why?
Use existing products as tests of requirements because they are yet another source of information about what is or is not desirable in the new product.

When?
Do this test when a draft of requirements is complete, though of course partial tests can be performed at any time during the requirements process. Indeed, when the old product is in daily use, it may be hard to avoid them, but don't assume that these on-the-fly tests are the same as a careful survey at the end.

How?
Do the following:

    1.   Compare the products to develop a list of possible missing functions in the new requirement.

2.   Interview users of the old product to develop a list of functions missing in the present system.

3.   Compare the old product with its original requirements to prepare a list of potential problems in developing the new product. Especially look for requirements that weren't implemented, or were implemented and then scrapped.

4.   Avoid the temptation to create a Swiss Army knife out of every product. Don't let features creep in without subjecting them to the full requirements process.

Who?

Try to find users who weren't involved in the original interviewing, in order to test your original sampling of users. If necessary, however, you can still gain much of value from the same users you interviewed initially. Also, try to find people who don't use the existing product, and find out why not.

24 MAKING AGREEMENTS

As discussed earlier, the decision tree pictures design as an exploration along the limbs of a tree of possibilities. The exploration starts at the root with the first vague statement of the problem, and it ends at a single leaf with the development of a single specific solution. The exploration proceeds by way of decisions, each of which is represented by a branch of the tree. As we take one fork or another, we reduce the ambiguity in the problem and get closer to the solution. But these decisions are simply dreams until we convert them to *agreements,* which is the subject of this chapter.

24.1 Where Decisions Come From

Most of this book has been addressed to the decisions nearest the root of the tree—at least they are the decisions that *should* be nearest the root. If these decisions have been well made, we can proceed with confidence and efficiency to design a product that will meet those requirements. If not, we will often need to go back down the tree to an earlier branch, or else settle for a less-than-ideal solution. We could say, then, that the purpose of all our requirements work is to put us on the right branch, and also to prevent backtracking.

24.1.1 Choices, assumptions, and impositions

By the time we've finished our requirements work, we've made hundreds or thousands of decisions. We've made many of those decisions directly and consciously. These we call *choices.* But not all decisions up to this point are choices, for some were made unconsciously by default, through bias, error, or lack of information. These decisions are called *assumptions.*

Other decisions were forced into the project—possibly by law, or custom, or higher authority—or perhaps they sneaked into the project without our noticing. In either case, these decisions were made *for* us by someone else, so we will call them *impositions*. Let's see how best to apply this knowledge using our elevator example.

Figure 24-1. The requirements work establishes a structure of strong limbs onto which the design, the implementation, and the product can grow. Without the requirements work, we may have all the information, but it may be cut up like a random pile of logs.

24.1.2 Elevator design decision examples

Here are some decisions in the design of the Elevator Information Device, each of which could be a choice, an assumption, or an imposition:

1.   There will be only one control/display panel per elevator car.

2.   The elevators will travel vertically only.

3.   All New York State elevator codes must be observed.

4.   No smoking or carrying of lighted tobacco products will be allowed on these elevators, and the system must be designed to enforce this provision.

5.   During a fire in the building, the elevators will be accessible to emergency personnel only.

6.   All but emergency power will be supplied by the building through the public utilities, to IEEE standards.

7.   The only source of emergency power available whenever the building power fails will be the backup power supplied by the Elevator Information Device.

8.   This backup power will be from storage devices such as batteries. It will not be generated.

9.   The backup power must be adequate to drive all necessary information device functions at their normal level of performance for at least four hours.

10. The elevators will be standard-sized cars for large buildings, ranging in floor area from 8 by 8 feet to 12 by 15 feet. Ceiling height will range from 8 to 12 feet.

11. The elevator cars will be rectangular in shape with flat, horizontally oriented floors and ceilings when in the normal operating position.

12. The elevator must be safe at all times.

13. The internal elevator temperature may vary from 60 to 80 degrees Fahrenheit.

14. At least one side and the floor of the elevator car will be totally available for controls, displays, and other necessary functions of the information device.

15. People will enter and leave the car from the front and/or rear through standard telescoping elevator doors.

16. People will observe the legal elevator load limits. There will be no stacking, pushing, shoving, or crushing of people.

17. The elevator speed will be limited to 1,440 feet per minute; the acceleration, to ± 4.4 feet per second squared.

18. The Elevator Information Device will be completely contained within or attached to the elevator—one device per car.

19. All information used or generated by the device will be available at the car. All information-sensing and -transmission functions will be the responsibility of other parties.

20. The car will be lighted at all times, except after four hours on emergency power.

21. Individuals who attempt to fool the system or make it look foolish (such as computer hackers) will be ignored.

22. No part of the device may intrude more than one-quarter inch into the elevator car.

24.1.3 Writing traceable requirements

Even with this partial list, you can see what controversies are avoided by making the jurisdiction of each of these choices explicit. For example, take this requirement:

2. The elevators will travel vertically only.

This requirement could also be an assumption—we never thought of an elevator as traveling horizontally, or even up an incline like the elevators on the Eiffel Tower. If we never thought of it, we could be limiting the market for our control system as design decisions depending on "vertical" become more and more hard-wired into the system.

Of course, any requirement that has been written down is presumably not an assumption. That's one of the best reasons for converting all choices to written agreements—anything that's not written down is automatically open for discussion at best, and is a poor, unilateral decision at worst.

If someone notices the vertical restriction, the discussion could be quite heated about where it came from. It could be an imposition required by law in certain states, but it could also be a choice if we consider the market for nonvertical elevators too small to justify the extra effort. Since it's important for the designers to know which it is, every requirement should be *traceable.*

For instance, if the requirement were an imposition, the restriction could be rewritten as

2. The elevators will travel vertically only *because none of the states in our marketing area legally allow nonvertical elevators.*

This informs the designers under what conditions it might be worth questioning this restriction, and perhaps gives them an estimate of how much it would be worth to allow for future nonvertical extension of the system.

On the other hand, if the restriction were a choice, it could be expressed as

2. The elevators will travel vertically only *because we don't see any appreciable market for nonvertical elevators at this time.*

Now the designers have slightly different information about what might happen in the future.

24.2 Where False Assumptions Come From

In the decision tree model, we can think of the requirements process as an orderly, reliable way of creating the trunk and the first major limb on which the design branch, the implementation twig, and finally the product leaf will grow. If our assumptions are not reliable, they will not form a reliable limb to support the design process.

24.2.1 Lack of valid information

Some assumptions are just false right from the start, though we are not aware that they are false. Under the assumption that the human body was immune to asbestos, builders used asbestos/concrete pipe and asbestos insulation sprayed on walls. This assumption caused thousands of deaths.

24.2.2 Invalidation over time

Some assumptions are true now, but gradually become false. When the telephone system in the United States was designed, calls were assumed to arrive more or less independently. But the telephone system was designed before the radio became widespread, and little by little, people all over the country began to be tied together by radio. On April 12, 1945, the announcement of Franklin Delano Roosevelt's death was broadcast over the radio. A large number of people heard the news at the same time and picked up their telephones to tell someone else. The entire telephone system was swamped and died that day, and so did the assumption of the independence of calls.

24.2.3 The turnpike effect

Economists sometimes make an assumption that they call the "perfect market." In a perfect market, selling a product doesn't affect the size or quality of the remaining market. In real life, of course, there are many "imperfect" markets, and some assumptions are rendered false by the product itself, once it comes into use.

For instance, certain load factors are assumed in designing a computer system, and the system is built with large capacity to handle the maximum assumed usage. But the old system's low capacity resulted in poor response, and so restrained the load. Now, when the new system's larger capacity removes the restraints, the load increases. Sometimes, the load changes so much that the new system is also swamped, but at a much higher level of traffic. This is called the "turnpike effect" or "parking lot effect," from two other common occurrences of the same phenomenon (Figure 24-2).

24.2.4 Requirements leakage

Assumptions can also be invalidated by *requirements leakage*. Whenever a product is used, new requirements are born. They may be requirements for more capacity as in the turnpike effect or requirements to "fix" problems—which may have been caused by mistakes in requirements, design, implementation, or training. They may be requirements for new functions, as the users master the existing function and now want to go where no one has gone before. If you look at any product a year after it reaches the users, you can be sure that literally hundreds of new "requirements" have leaked in (see Figure 24-3).

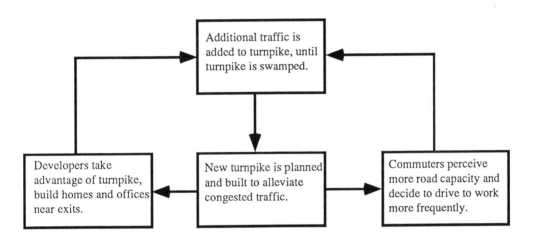

Figure 24-2. **The turnpike effect, or parking lot effect, occurs when the introduction of a new facility with larger capacity ultimately results in swamping by calling forth more load.**

24.3 Converting Decisions to Agreements

The purpose of requirements work is to guide future actions in developing a product. But even if all the decisions have been made correctly, they may not suffice to guide future actions correctly. In particular, assumptions and impositions create conditions not under the developers' direct control, conditions that determine the success or failure of the project. These assumptions and impositions represent *risks* to timely and efficient project completion.

Because impositions can be seen as intrusions, some of the implementors may resent them, or at least disagree with them. They may refuse to conform to the impositions, yet every nonconforming action drops the project off the correct branch of the design tree.

Because assumptions may be implicit, some people probably won't really understand them. Similarly, others will misunderstand choices that are made but not communicated. Indeed, if choices are not communicated, some people may simply be unaware of their existence. So, even if participants' intentions are blameless, they may fall off the correct design branch out of ignorance.

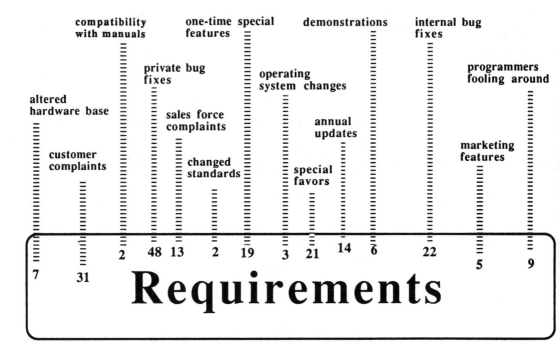

Figure 24-3. Requirements leakage occurs when new requirements are introduced by other than the official requirements process. In one software organization, 202 leaks were documented for a single product, in 6 weeks, from 14 different sources.

In short, much of the requirements work will be negated if choices, impositions, and assumptions are not both understood and accepted by everyone involved. Thus, before moving out of the requirements phase into the rest of the design process, all parties must understand and accept their responsibilities. Otherwise, customers will be disappointed when the product is delivered. To ensure understanding and acceptance, you must attempt to convert every choice, imposition, and assumption into explicit, documented *agreements*.

24.4 Helpful Hints and Variations

- We like to have people actually sign their names to signify agreement. Some people object that this is too formal, but we've found that without a signature, a document is just another piece of paper with little value. The signature adds value.

- Think of signing as yet another test. If people hesitate to sign, don't try to pressure them into signing. Instead, work with them to discover what's holding them back. You'll almost always find another assumption that's worth documenting.

- "Teams may work from sun to sun, but requirements work is never done." Some definition of assumptions is always left to the designers, builders, and users. What you can do, however, is to *document the meta-assumption that certain assumptions are being left to others to define.*

- One way to document assumptions left to others is to create *safety factors*, a traditional way for engineers to say, "We don't know what to assume here; therefore we're allowing a large margin for error—the error in our knowledge of assumptions." Take your lead from the engineers and express safety factors in your own agreements by saying simply, "We don't know what to assume here, so we're assuming a wide range. This may lead to overdesign, so watch for opportunities to narrow this assumption."

- The safety factor image shows that what underlies the attempt to get agreement is the concept of *risk*. When you try to convert assumptions to agreements, you are really trying to reduce risk. As you go through this process, you don't have to think of it as pure documentation. Ideally, you will notice areas where you can transform some of these assumptions into *actions that will reduce the risk.* For example, you may agree to a process for seeking what's missing, or for buying insurance against failure, or for involving someone with special resources in the development.

24.5 Summary

Why?
Elevate all assumptions to consciousness so that you can control them as the development process continues.

When?
Just before you believe you are finished, test your assumption that you are done by getting explicit, written agreements.

How?
To move toward agreement, take the following steps:

1. Write down every assumption.

2. Trace each assumption to its source: Is it a choice, an imposition, or an agreement?

3. Add information to the assumption that identifies the source.

4. Get participants to sign their names to each written document.

5. Look for opportunities to agree to actions that will reduce risk caused by assumptions.

Who?

Anyone left out of the agreement phase will show up later and make trouble, so include everyone whose later objections could be troublesome.

25   ENDING

The Book of Life begins with a man and woman in a garden.
It ends with Revelations.

> —Oscar Wilde, *A Woman of No Importance*

25.1 The Fear of Ending

The requirements process begins with ambiguity. It ends not with revelations, but with agreement. More important, it ends.

But *how* does it end? At times, the requirements process seems like Oscar Wilde when he remarked, "I was working on the proof of one of my poems all the morning and took out a comma. In the afternoon, I put it back again."

A certain percentage, far too large, of development efforts never emerge from the requirements phase. After two, or five, or even ten years, you can dip into the ongoing requirements process and watch them take out a comma in the morning and put it back again in the afternoon. Far better that they should be killed during requirements than allowed to live such a lingering death.

Paradoxically, it is the attempt to finish the requirements work that creates this endless living death. The requirements *phase* ends with agreement, but the requirements *work* never ends until the product is finished. There simply comes a moment when you decide you have enough agreement to risk moving on into the design phase.

25.2 The Courage to End It All

Nobody can tell you just when to step off the cliff. It's simply a matter of courage. Whenever an act requires courage, you can be sure that people will invent ways of reducing the courage needed. The requirements process is no different, and several inventions are available to diminish the courage needed to end it all.

25.2.1 Automatic design and development

One of the persistent "inventions" to substitute for courage is some form of automatic design and/or development.

In automatic development, the finished requirements are the input to an automatic process, the output of which is the finished product. There are, today, a few simple products that can be produced in approximately this way. For instance, certain optical lenses can be manufactured automatically starting from a statement of requirements.

In terms of the decision tree, automatic development is like a tree with a trunk, one limb, no branches or twigs, and a single leaf (Figure 25-1). With such a system, finishing requirements is not a problem. When you think you might be finished, you press the button and take a look at the product that emerges. If it's not right, then you weren't finished with the requirements process.

Figure 25-1. Automatic development is like a tree with a trunk, one limb, no branches or twigs, and a single leaf.

No wonder this appealing dream keeps recurring. It's exactly like the age-old story of the genie in the bottle, with no limit on the number of wishes. For those few products where such a process is in place, the only advantage of a careful requirements process is to save the time and money of wasted trial products. If trial products are cheap, then we can be quite casual about finishing requirements work.

25.2.2 Hacking

Automatic development is, in effect, nothing but requirements work—all trunk and limb. At the other end of the spectrum is *hacking*, development work with no explicit requirements work. When we hack, we build something, try it out, modify it, and try it again. When we like what we have, we stop. In terms of the decision tree model, hacking is not a tree at all, but a bush: all branches, twigs, and leaves, with no trunk or major limb (Figure 25-2).

Figure 25-2. Hacking is not a tree at all, but a bush—all branches, twigs, and leaves, with no trunk or major limb.

Pure hacking eliminates the problem of ending requirements work, because there is no requirements work. On the other hand, we could conceive of hacking as pure requirements work—each hack is a way of finding out what we really want.

Almost any real project, no matter how well planned and managed, contains a certain amount of hacking, because the real world always plays tricks on our assumptions. People who abhor the idea of hacking may try to create a perfect requirements process. They are the very people who create "living death" requirements processes.

25.2.3 Freezing requirements

Paradoxically, hacking and automatic development are exactly the same process from the point of view of requirements. They are the same because they do not distinguish requirements work from development. Most real-world product development falls somewhere between pure hacking and automatic development, and that is why we have to wrestle with the problem of ending.

Even when we have every requirement written down in the form of an agreement, we cannot consider the requirements process to be ended. We know that those agreements will have to change because in the real world, assumptions will change. Some people have tried to combat their fear of changing assumptions by imposing a freeze on requirements. They move into the design phase with the declaration, *No changes to requirements will be allowed.*

Those of us who understand the nature of the real world will readily understand why a freeze simply cannot work. We know of only one product development when it was even possible to enforce the freeze. In that instance, a software services company took a contract to develop an inventory application for a manufacturer. The requirements were frozen by being made part of the contract, and eighteen months later, the application was delivered. Although it met all the contracted requirements, the application was rejected.

This frozen product was totally unusable because in eighteen months it had become totally removed from what was really required in the here-and-now. The customer refused to pay, and the software services company threatened legal action. The customer pointed out how embarrassing it would be to a professional software company to have its "freeze fantasy" exposed in a public courtroom. Eventually, the two parties sat down to negotiate, and the customer paid about one-fourth of the software firm's expenses. At the end of this bellicose negotiation, someone pointed out that with far less time negotiating, they could have renegotiated the requirements as they went along, and both parties would have been happy.

25.2.4 The renegotiation process

The freeze idea is just a fantasy, designed to help us cope with our fear of closure. But we cannot fearlessly close the requirements phase unless we know that there is some renegotiation process in place. That's why agreeing to a renegotiation process is the last step in the requirements process.

Working out the renegotiation process is also a final test of the requirements process itself. If the requirements process has been well done, the foundation has been laid for the renegotiation process. The people have all been identified, and they know how to work well together. They have mastered all the techniques they will need in renegotiation, because they are the same techniques the participants needed to reach agreement in the first place.

25.2.5 *The fear of making assumptions explicit*

The agreement about renegotiation must, of course, be written down, and this act itself may strike fear in some hearts. Another way to avoid ending requirements is to avoid written agreements at the end of the requirements phase. Some designers are afraid of ending requirements because the explicit agreements would make certain assumptions explicit. An example may serve to make this surprising observation more understandable.

While working with the highway department of a certain state, we encountered the problem of what to do about a particularly dangerous curve on one of the state highways (see Figure 25-3). In an average year, about six motorists missed the curve and went to their death over a cliff. Because it was a scenic highway, it was neither practical nor desirable to eliminate the curve, but highway design principles indicated that a much heavier barrier would prevent wayward cars from going over the cliff.

Figure 25-3. What should be done about this dangerous curve?

Building the barrier would seem like an obvious decision, but there was another factor to consider. Perhaps once every three years, with a heavy barrier in place, one of these wayward cars would bounce off the barrier into a head-on collision with an oncoming vehicle. The collision would likely be fatal for all involved.

Now, on average, the number of people killed with the barrier in place would be perhaps one-fifth of those killed without the barrier, but the highway designers had to think of another factor. When a solo driver goes over the cliff, the newspapers will probably blame the driver. But what if a drunk driver bounces off the barrier and kills a family of seven who just happened to be driving in the wrong place at the wrong time? The headlines would shout about how the barrier caused the deaths of innocent people, and the editorials would scream for heads to roll at the highway department.

So, it's no wonder that the highway engineers didn't want anything documented about their decision not to build the barrier. They were making life-and-death decisions in a way that covered their butts, and they could protect themselves by taking the position, "If I never thought about it, I'm not responsible for overlooking it in the design." And, if it was never written down, who could say they had thought of it?

25.3 The Courage to Be Inadequate

Most engineers and designers react to this story by citing similar stories of their own to prove that, yes, indeed, there are decisions that it's better not to write down. We believe, however, that this kind of pretense is an abuse of professional power, an abuse that is not necessary if we remember the proper role of the requirements process.

It's not for the designers to decide what is wanted, but only to assist the customers in discovering what they want. The highway designers should have documented the two sides of the issue, then gone to the elected authorities for resolution of this open requirements question. With guidance from those charged with such responsibilities, the engineers could have designed an appropriate solution.

But suppose the politicians came back with an impossible requirement, such as

The highway curve must be redesigned so that there will be no fatalities in five years.

Then the engineers would simply go back to their customers and state that they knew of no solution that fit this requirement, except perhaps for a barricade that prevented cars from using the highway. Yes, they might lose their jobs, but that's what it means to be a professional—never to promise what you know in advance you can't deliver.

The purpose of requirements work is to avoid making mistakes, and to do a complete job. In the end, however, you can't avoid all mistakes, and you can't be omniscient. If you can't risk being wrong, if you can't risk being inadequate to the task you've taken on, you will never succeed in requirements work. If you want the reward, you will have to take the risk.

25.4 Helpful Hints and Variations

- If you want to be a professional designer, consider first the task of designing yourself, especially so you have the courage to act professionally. Don't spend all your time working on the technical side of requirements work, or on facilitating what other people do. Spend at least half of your time working on yourself.*

25.5 Summary

Why?
Pay special attention to ending the requirements process because fears of imperfection can lead you into endless cycles.

When?
You know when you have reached the end when you have all necessary agreements in hand. Fear of ending has nothing to do with defining the end, except that you will never end if you don't face your fear.

Who?
All the people involved in the work must agree that you have reached an end, otherwise they will never stop making changes. You know they have agreed when they have committed to a renegotiation process.

How?
Ending requirements work is much like ending a book—especially a book on as open a subject as requirements definition. There comes a moment when you still have more to say, much more, but you find yourself taking out commas in the morning and putting them back in the afternoon. At that point, you simply accept the risk of leaving some ideas imperfect and some revelations invisible. You check that you haven't lost any written material, you gather your courage, and you stop.

---

*For some ideas on self-development as a technical leader, see, for example, Gerald M. Weinberg, *Becoming a Technical Leader* (New York: Dorset House Publishing, 1986).

BIBLIOGRAPHY

Alexander, Christopher. *Notes on the Synthesis of Form.* Cambridge, Mass.: Harvard University Press, 1979.

> *Design is examined as a generic activity practiced along an unself-conscious/self-conscious continuum. The design process is viewed as one that attempts to find a fitness between two entities: the form in question and its context.*

Ashby, W. Ross. *An Introduction to Cybernetics.* London: Chapman and Hall, 1964.

> *Contains the early frameworks of cybernetic or feedback systems; it is quite useful for exploring the design process as a feedback system with the client and designer as information channels.*

Boehm, Barry W. *Software Engineering Economics.* Englewood Cliffs, N.J.: Prentice-Hall, 1981.

> *Provides raw numbers collected from a large number of completed software projects and suggests various metrics.*

Brooks, Frederick P. *The Mythical Man-Month.* Reading, Mass.: Addison-Wesley, 1982.

> *An extremely informative report of experiences gleaned during the design and development of what was, at the time, one of the most complex systems created by mankind—OS/360. There are many lessons here for both large and small systems design.*

Buzan, Tony. *Using Both Sides of Your Brain.* New York: E.P. Dutton, 1976.

> *The original, and probably still the best, reference on the subject of mind maps.*

Carroll, Lewis, and Martin Gardner. *The Annotated Snark.* New York: Simon and Schuster, 1962.

This is Lewis Carroll's Hunting of the Snark, *with extensive annotations by Martin Gardner. Don says that this book doesn't really have anything to do with design (or does it?) but thinks everyone should read it anyway. Jerry thinks it's a book about how projects are done without understandable requirements.*

Case, Albert F., Jr. *Information Systems Development: Principles of Computer-Aided Software Engineering.* Englewood Cliffs, N.J.: Prentice-Hall, 1986.

An excellent survey of the case for CASE *through 1986, as well as a good introduction to the ideas of using computers to aid computer software development projects.*

Conte, S.D., H.E. Dunsmore, and V.Y. Shen, *Software Engineering Metrics and Models.* Menlo Park, Calif.: Benjamin-Cummings Publishing Co., 1986.

A formal statistical treatment of software engineering metrics. Of use primarily in estimating software complexity, productivity, and cost.

Curtis, Bill, ed. *Tutorial: Human Factors in Software Development.* Los Angeles: IEEE Computer Society, 1981.

A valuable collection that contains a reprint of Weinberg and Schulman, as well as other seminal articles on software development in general, and requirements in particular.

DeMarco, Tom. *Structured Analysis and System Specification.* Englewood Cliffs, N.J.: Prentice-Hall, 1978.

The classic book on using structured English to specify requirements for software development projects. It's well-written, and well worth reading.

_____. *Controlling Software Projects.* Englewood Cliffs, N.J.: Prentice-Hall, 1982.

Contains useful information for both managers and developers involved in the planning, organizing, and time- and cost-estimating of large software development projects.

Deming, W. Edwards. *Out of the Crisis.* Cambridge, Mass.: MIT Center for Advanced Engineering Study, 1986.

Deming has influenced many people on the relationship of requirements work to quality, promoting such topics as the measurability of preferences.

Doyle, Michael, and David Straus. *How to Make Meetings Work: The New Interaction Method.* Chicago: Playboy Press, 1976.

A small, readable, inexpensive book to hand out to people who haven't had much experience with, or hope for, productive meetings.

Edwards, Betty. *Drawing on the Artist Within.* New York: Simon and Schuster, 1986.

Based on recent research in dual brain function, this work provides a means for enhancing creative activity within oneself by taking the reader on a trip of visual experiences and other right-brain exercises.

Freedman, Daniel P., and Gerald M. Weinberg. *Handbook of Walkthroughs, Inspections, and Technical Reviews,* 3rd ed. Chicago: Scott, Foresman and Co., 1983.

Contains a wealth of information on technical reviews, including checking test cases, a list of dangerous words to look for, heuristics for inspecting documents, and extensive checklists for reviewing requirements. It's in a question/answer format for easy access to just the information you need.

Gause, Donald C., and Gerald M. Weinberg. *Are Your Lights On? How to Know What the Problem Really Is,* 2nd ed. New York: Dorset House Publishing, 1989.

In this book, we have addressed many difficulties encountered in the art of problem definition. The book is light enough to serve as a catalyst between systems designers—the problem solvers—and clients and end users—the people with the problems.

Glegg, G.L. *The Design of Design.* Cambridge, England: Cambridge University Press, 1969.

_____ . *The Science of Design.* Cambridge, England: Cambridge University Press, 1973.

_____ . *The Selection of Design.* Cambridge, England: Cambridge University Press, 1973.

_____ . *The Development of Design.* Cambridge, England: Cambridge University Press, 1981.

Gregory, S.A., ed. *The Design Method.* London: Butterworth, 1966.

The four books by Glegg and this one edited by Gregory deal with methods for solving problems associated with the design of complex systems.

Grosser, M. *Gossamer Odyssey*. Boston: Houghton Mifflin, 1981.

Gives an interesting description of how designers have combined the self-propelled requirement with air transportation.

Halprin, Lawrence. *The RSVP Cycles: Creative Processes in the Human Environment.* New York: George Braziller, Inc., 1969.

The author, an environmental designer and planner by profession, introduces a design approach to the modeling of creative production in the arts. He then extends the approach to areas of traditional design as a generic design tool.

Hanks, Kurt, Larry Belliston, and Dave Edwards. *Design Yourself.* Los Altos, Calif.: William Kaufmann, Inc., 1977.

This workbook playfully introduces the reader to many proven means of breaking mind-sets in the quest for more creative solutions and problem-solving approaches.

Hatley, Derek J., and Imtiaz A. Pirbhai. *Strategies for Real-Time System Specification.* New York: Dorset House Publishing, 1987.

The title of this book is just about the only ambiguous part. It's not about how to specify a system in real time, but how to specify systems that will have real-time requirements. It's outstandingly thorough at covering the formal parts of the process, which makes it an excellent pair with Exploring Requirements *and its human orientation.*

Jones, David E.H. *The Inventions of Daedalus.* Oxford: W.H. Freeman and Co., 1982.

Contains 102 well-illustrated, outrageous ideas for products including audible vertigo, silenced and subverted sound, tired light, smell amplification, magnetic fur, and non-Newtonian trousers. Written by a British scientist, it is an excellent starting point for any design project requiring large amounts of innovation.

Jones, J. Christopher. *Design Methods.* New York: Wiley-Interscience, 1980.

Discusses many specific solution approaches to design problems.

Keirsey, David, and Marilyn Bates. *Please Understand Me: Character & Temperament Types*, 4th ed. Del Mar, Calif.: Prometheus Nemesis Book Co., 1984.

Still the best reference on the Myers-Briggs Type Indicator, a method of understanding and dealing successfully with the various personalities you encounter in life, and in requirements work.

Koberg, Don, and Jim Bagnall. *The Revised All New Universal Traveller.* Los Altos, Calif.: William Kaufmann, Inc., 1981.

To quote the authors, "Design is the process of making dreams come true." Fantasy, ideation, and creativity are introduced as an important part of the design and problem-solving process.

Marca, David A., and Charles L. McGowan. *SADT™ Structured Analysis and Design Technique.* New York: McGraw-Hill, 1988.

The authoritative coverage of the subject of information systems analysis using the techniques pioneered by Doug Ross, who wrote the Foreword.

Martin, James, and Carma McClure. *Diagramming Techniques for Analysis and Programming.* Englewood Cliffs, N.J.: Prentice-Hall, 1985.

The most comprehensive survey of notations for information systems analysis and design, including decision trees and tables, flow charts, HIPO (hierarchic input-process-output) charts, Jackson diagrams, Nassi-Shneiderman charts, state transition diagrams, structure charts, and Warnier-Orr diagrams.

McKim, Robert H. *Thinking Visually.* Belmont, Calif.: Wadsworth, Inc., 1980.

Approaches are introduced to aid the problem-solver in dealing with problems in nontraditional, visually oriented ways. Concepts involving ambidextrous thinking, relaxed attention, pattern-seeking, autonomous imagery, and idea-sketching are discussed.

McMenamin, Stephen M., and John F. Palmer. *Essential Systems Analysis.* Englewood Cliffs, N.J.: Prentice-Hall, 1984.

An excellent all-round introduction to information systems analysis, especially good on the subject of how to analyze existing systems.

Osgood, Charles E., George J. Suci, and Percy H. Tannenbaum. *The Measurement of Meaning.* Urbana, Il.: University of Illinois Press, 1957.

The source of the idea of a semantic differential, which is the progenitor of our user satisfaction test.

Perry, William E. *A Structured Approach to Systems Testing.* Wellesley, Mass.: QED Information Sciences, 1983.

A comprehensive book on testing information systems that contains a useful chapter on testing during the requirements phase.

Petroski, Henry. *To Engineer Is Human.* New York: St. Martin's Press, 1985.

Deals primarily with mechanical and structural systems and provides excellent examples of systems failures due to misinformation or lack of information.

Satir, Virginia. *Making Contact.* Berkeley, Calif.: Celestial Arts, 1976.

Virginia Satir knew more about how people could work together than anybody else we can imagine. This little book is about the way we cope with stressful situations and present ourselves to others. It's a good place to start the adventure of getting to know her work, and yourself.

Spradley, James P. *Participant Observation.* New York: Holt, Rinehart & Winston, 1980.

An outstanding introduction to participant observation—and a book that requires no background in the social sciences.

Weinberg, Gerald M. *The Secrets of Consulting.* New York: Dorset House Publishing, 1985.

A guide for anyone who offers advice at the request of other people. It contains, among other topics, a complete explanation of the Orange Juice Test.

_____ . *Becoming a Technical Leader.* New York: Dorset House Publishing, 1986.

Full of ideas on self-development as a technical leader, including how to lead product development work.

_____ . *Rethinking Systems Analysis & Design.* New York: Dorset House Publishing, 1988.

Takes off where introductory books on systems analysis leave off. It contains information on black boxes, a complete discussion of optimitis and its cure, a fuller development of the Railroad Paradox, complete discussion of Wiggle Charts, and many practical ideas about observing and interviewing.

_____ , and E.L. Schulman. "Goals and Performance in Computer Programming," *Human Factors*, Vol. 16, No. 1 (1974), pp. 70-77.

_____ , and Daniela Weinberg. *General Principles of Systems Design.* New York: Dorset House Publishing, 1988.

Shows how many of the deep principles of systems design can be derived from the necessity for survival.

Williams, C. *Craftsmen of Necessity.* New York: Random House, 1972.

Many examples of indigenous design are introduced. This unself-conscious design process involves the end user as the designer. The design and implementation, indistinguishable from one another, are accomplished in an evolutionary, trial-and-error process. The resultant product, incorporating vast amounts of environmental information, is likely to be highly successful.

Index

Agenda, for meetings, 87-88, 111
Agreements, 268-76, 277
    for renegotiation, 280-83
Alexander, C., 285
Ambiguity:
    assumptions and, 16-18, 35
    cost of, 17-18, 21
    design process, 220-21
    final-product, 220
    metric, 24, 217-24
    name and, 54-58, 128-35
    notation and, 8-13
    problem statement and, 14-21, 27-33,
    92-103, 217-21
    -reducing techniques, 36-45, 92-103,
    247
    sources of, 22-33, 45, 92ff., 157,
    163-65, 219-21
    words and, 16-17, 93-103
Ambiguity poll, 29, 93, 94, 102, 223-24
    star example, 93-94
    transportation device example, 42,
    217-22
Amendment procedure, 86, 143
Aristotle, 260
Asbestos example, 18
Ashby, W., 285
Assumptions, 268, 271, 274-75
    Can-Exist, 55-58
    existence, 149
    explicit, 281
    false, 18, 35, 271-73
    negotiating and, 143
Attributes, 150, 158, 161-70

ambiguity and, 163-65
details, 163-66, 170
elevator example, 162-67, 172, 187-93
function and, 166-67, 169-70
must, want, ignore, 167-171
optional condition, 187
preferences and, 186-200
user satisfaction test and, 248
wish list, 161-66
Automated credit card verification ex-
    ample, 158-59
Automated development, 3-11, 278-79

Backronym, 134
Bagnall, J., 289
Bates, M., 146, 288
Bell, A., 260-61
Belliston, L., 288
Berra, Y., 223
Bipolar adjective pairs, 239
Black box testing, 249-52, 257, 258
BLT Design, 47-48, 68, 147-49, 161ff.,
    174-77, 186ff., 215-16
Boehm, B., 17-18, 285
Brainblizzard, 109-11
Braindrawing, 123
Brainstorm, 109, 111-19
    attributes list, 158, 161-63, 170
    functions list, 150-52, 156-58
    user constituencies, 73, 78
Brainwriting, 117-18, 123
Broadcasting technique, 77
Brooks, F., 285
Buzan, T., 285

CAD/CAM example, 241-42
Can-Exist assumption, 55-58
Carroll, L., 120, 125-26, 128, 285, 286
Case, A., 286
CASE and CAD, 3-11, 286
CASEware, 78
Cheap Chalk Corp., 147-48, 149, 161,
    164, 168, 172, 186
Choices, 268, 275
Client: see Customer, Designer, User
Clustering heuristic:
    for star example, 26-33
    transportation device example, 42,
    221-22
Colonel Stoopnagle, 230
Communication problems, 13, 127
    project names and, 54-58, 128-35
Computer programming experiment, 1-2
Conflicts, 136-46
    and expectations, 208ff.
    measurability, 184
    personality, 137-38, 141-43
    political, 144, 237
Constraints, 171-85
    of elevator example, 171-72, 174,
    178-79, 187-93
    features and, 265
    must attributes and, 171, 185
    psychology of, 180-82
    of Superchalk, 171-79, 186-88, 192
    testing, 174-77, 185, 255-56
    versus preferences, 189-200
Conte, S., 286
Context-free questions, 59-67, 68
    design process and, 59-60, 64, 67
Convergent design, 22-25
    processes slide, 101
Costs, 17-18, 21, 144, 285
    estimating, 286
    preferences and, 187
    trade-offs and, 195-200
    what's-it-worth? graph, 192-93
Cowboy example, 60-61
Curtis, B., 1, 286
Customers, 11
    identifying, 68-79, 161
    needs versus wants, xv, 44-45, 201-13

preferences of, 186-200
product approval of, 149, 185, 259
reviews and, 227-37
versus users, 68-69
(see also Attributes, Designer, Function, User)

Decision tree model, 34-45, 268-75, 289
    context-free questions, 61
    example, 36-44
de Lavigne, K., 12, 21, 159, 212, 222, 247
DeMarco, T., 286
Deming, W., 21, 286
Designer:
    -client conversation, 11, 13, 17, 94,
    211, 282, 285
    decision tree and, 36-44, 268ff.
    functions and, 157
    measuring satisfaction and, 205,
    212-13, 238-48
    preferences and, 191-92, 199
    test cases and, 254-55
Design model, 34ff., 219
Design process, 59-60, 64, 67, 285, 291
    ambiguity, 220-21
Development time, 144
    (see also Schedule constraint)
Direct question, 34ff., 44, 67
    (see also Interviewing techniques)
Doctrine of Reasonable Use, 211-12
Documenting, xvi, 212
    assumptions, 275
    limitations, 210, 213
    reviews and, 230-33, 236
    test cases, 256-57, 259
    (see also Maps)
Do Not Disturb Project, 109-11, 128-29,
    225-27
    user satisfaction test for, 239-40
Doyle, M., 89, 287
Drinking glass example, 121, 123-25
Dunsmore, H., 286

Edwards, B., 287
Edwards, D., 288
Einstein, A., 180
Eisenhower, D., xvi

Electronic mail, 118
Elevator Information Device Project,
    55-57, 262
    attributes of, 162-67, 172, 187-93
    choices, impositions, and assump-
    tions, 268ff.
    constraints, 171-72, 174, 178-79, 187-93
    context-free questions, 59-62
    expectation list, 204-10
    functions of, 149-57
    metaquestions and, 62-64
    preferences and, 187-93
    solution space of, 174, 179
    testing, 250-56
    user constituencies and, 72-73
Environment:
    physical, 16, 291
    of safety (meetings), 84-89, 91
    shared expertise, 11
    social and cultural, 16
Estimation, 221-24, 286
Evident function, 152-55, 158, 159
Examples:
    asbestos, 18
    automated credit card verification,
    158-59
    drinking glass, 121, 123-25
    elevator: see Elevator Information
    Device Project
    Ford Pinto, 18
    FORTRAN compiler, 246-47
    furniture design, 129-34
    highway, 281-82
    Holiday Inn, 202-203, 211
    iced tea glass, 123-25
    lecture, 22-33
    Legionnaires' Disease, 25, 31
    letter writing, 6-10, 121-23
    Lewis Carroll, 120, 125-26
    nursery rhyme, 94-103
    racing car, 195-98
    radar/power supply, 202-203, 210
    Wright brothers, 53-54
Exhaustive participation, 74-75
Expectations, 201-213
    for elevator device, 204-10
    limitation process, 204-208, 211-13

list, 204ff.
    of Superchalk, 201, 205
Explicit solution, 64
Exploring requirements, xvi, 12-13, 80,
    105-107, 127, 136
    decision tree and, 34-45, 268
    elevator example of, 55-57
    model of, 20-21, 105-106
    picture of, 19

Facilitators, 112-15, 136-46
    professional, 144-45, 234
Feasibility study, 58
Final-product ambiguity, 220
FitzGerald, E., 12
Ford Pinto example, 18
Formal methodology, 4-6, 9
FORTRAN compiler example, 246-47
Freedman, D., 95, 183, 225, 234, 251, 287
Frill function, 152-55, 159, 191
    "Get It If You Can" list, 157, 160
Front-end effort, 36
Function, 16
    attributes and, 166-67, 169-70
    defining, 149-50
    of elevator example, 149-57
    evident, hidden, frill, 152-55, 157-59,
    160, 191
    existence, 149, 159
    "Get It If You Can" list, 157, 160
    list, 150-52, 156-68
    missing, 262-63
    opaque, 158-59
    overlooked, 155-56, 159
    of Superchalk, 149
    testing, 150, 249-59
    translucent, 158
    versus features, 264-67
Furniture design example, 129-34

Gane-Sarson approach, 6
Gardner, M., 285
Gause, D., 22-33, 49, 95, 287
"Get It If You Can" list, 157, 160
Glegg, G., 287
Global properties, 59, 64, 67
Gregory, S., 287

Grosser, M., 42, 288
Guaranteed Cockroach Killer, 3-4, 13, 127

Hacking, 279
Halprin, L., 288
Hanks, K., 288
Hatley, D., 288
Heuristics:
    clustering, 26-33, 42, 221
    "Mary conned the trader," 95-97, 103
    "Mary had a little lamb," 94-103
    memorization, 92-94, 102
    naming, 56-57, 132-35, 138
    recall, 32
    user-inclusion, 72-74
Hidden function, 152-56, 158-59
Highway example, 281-82
Histogram, for star example, 27, 30, 93
Holiday Inn example, 202-203, 211
Human limitations, 29, 45, 201-204

Iced tea glass example, 123-25
Idea-generation methods, 105-106, 123, 134
    meetings and, 109-19
    solution space and, 176-77
Idea-reduction methods, 105-106, 115-18
Implicit:
    attribute, 168
    constraint, 180
    function, 155-56, 159
    preference, 199
    solution, 64, 156-57, 160, 168-69,
        208-209
Impositions, 268-69, 273, 275
Indispensable participants, 138-39
Initiation process, 49-50
Inspection, 95, 234, 235, 287
Intercluster variation, 28-29, 33, 102
Interpretation errors, 27-29, 32-33, 102
Interviewing techniques, 34-45, 57, 290
    to identify functions, 261-64, 267
    to limit expectations, 204
    videotaping, 212
Intracluster variation, 28-29, 32, 102

Jackson approach, 6, 289
Johns-Manville Corp., 18

Jones, D., 288
Jones, J., 288
Keirsey, D., 146, 288
Koberg, D., 289

Language, 11
    programming, 183-85
    (see also Notation)
Lawyer example, 189-90
Lecture example, 22-33
Legionnaires' Disease example, 25, 31
Letter writing example, 6-10
    Wiggle Charts of, 121-23
Lewis Carroll example, 120, 125-26
Limitations, 204-13
    (see also Constraints)

Management level conflicts, 139-40
Maps:
    of letter writing example, 6-10, 121-23
    mind, 285
    notation, 6-13, 127
    ocean, 126
    right-brain methods and, 120-27
    tools, 120-23
Marca, D., 289
Martin, J., 289
"Mary conned the trader" heuristic,
    95-97, 103
"Mary had a little lamb" heuristic,
    94-98, 101-103
McClure, C., 289
McGowan, C., 289
McKim, R., 289
McMenamin, S., 260, 289
Measureability, 184, 188-89
    of constraints, 184
    meetings and, 83-84
    of preferences, 188-89, 199-200, 286
    of requirements document, 230
    of satisfaction, 205, 212-13, 238-48
Meetings, 80-91, 287
    example, 80-83
    facilitator and, 90, 145-46
    idea-generation, 109-19
    participation in, 84-91
    (see also Technical reviews)

Memorization heuristic, 92-94, 102
Messy problem, 5-6
Metaphorical thinking, 52, 57
    (*see also* Simile)
Metaquestions, 62-64
    for elevator example, 62-64
    examples of, 65-67
Methodology, 3, 183
    automated, 3-11
Metrics, 188-89, 199-200, 285, 286
    ambiguity, 24, 217-24
Minch, E., 11, 212, 223
Mockup, 53-55, 57
Model-in-the-lobby technique, 75-76
Must attribute, 167-70, 171, 185

Name (project), 54-58, 128-35
    elevator example, 57
    furniture design example, 129-34
Naming heuristic, 56-57, 132-35
    face-to-face session, 134, 138
Negotiation, 143
    of constraints, 176, 184
    expectations, 208
    facilitators and, 136, 143
    renegotiation and, 280, 283
    solution space and, 185
No parking example, 189-90
Norm, 53-54, 57
    existing product as, 260-61
Notation:
    maps and, 6-13, 127
    Swiss example, 11
Nursery rhyme example, 94-103

Observational error, 27-29, 32-33, 102
Opaque function, 158-59
Optimitis, 194-98, 290
Optimization statements, 159
Orange Juice Test, 199, 290
Osborne, A., 109, 111
Osgood, C., 238, 289
Overconstraint, 179-80, 182, 185, 199,
    255-56, 259
    (*see also* Constraints)

Palmer, J., 260, 289

Pareto Optimization, 71
Participant observation, 78, 290
Participation, xv, 33, 58, 213, 259, 276
    exhaustive, 74-75
    facilitators and, 112-15, 136-46
    identifying, 68-79
    indispensable people, 138-39
    in meetings, 84-91, 119
    naming heuristic and, 134-35
    notation and, 6-13, 127
    in requirements process, xv, 74-79, 211
    sample, 74-77
    in technical reviews, 227-37
Payroll package example, 44
Perry, W., 290
Personality clash, 137-38
Personality differences, 141-43
Petroski, H., 290
Pirbhai, I., 288
Post-It™ note example, 51
Preferences, 186-200, 286
    for elevator example, 187-93
    for Superchalk, 186-88, 192
    versus constraints, 189-200
Problem, defined, 49-50
Problem solution:
    acceptable, 149, 185ff., 287
    decision tree and, 34-45
    idea, 50-51
    implied, 64, 156-57, 160, 208-209
    messy, 5-6
    phase, 24
    solution space and, 173-85, 186ff.
    structure examples, 14-17
    transportation device example, 42-43
Problem statement, 1, 160
    ambiguity, 14-21, 27-33, 92-103, 217-21
    house (structure) example, 14-17
    star focus slide example, 25, 92-94,
    97-101
    transportation device example, 36,
    217
    universal furniture design, 133
Process questions, 59-60, 65
Product questions, 61-62, 65
Programming team performance, 1-2
Project review, 228-29

Prototype, 77
    as satisfaction test, 246-47
Puppy simile, 52, 53

Questions: *see* Interviewing techniques

Racing car example, 195-98
Radar/power supply example, 202-203, 210
Railroad Paradox, 69-70, 290
Recall errors, 27-29, 32-33, 92, 102
Recall heuristic, 32
Reframing, 142-43
    constraints into preferences, 194-200
Reports, 230-33, 236
Requirements:
    familiarization problem, 8-9
    leakage, 272-74
    maps and, 9-13, 127
    missing, 16
    phase, 8, 24, 60, 144, 277
    statement, defined, 1
    users and, 9, 11
    (*see also* Ambiguity, Attribute, Exploring, Function)
Requirements document, 1, 32
    freezing, 280
    reviewing, 95, 227-37
    for Superchalk, 215-16
    testing, 261-64
Requirements process, 8-13
    black box techniques, 249-52, 257, 258
    consulting services for, 78
    context-free questions and, 59-68
    decision-making, 268-76
    development of constraints, 171-85
    identifying preferences, 198
    issues list, example of, 231
    project name, 54-58, 128-35
    starting points, 49-58, 260, 277, 288
    technical reviews in, 199, 225-37
    user participation in, xv, 74-79, 211
    user satisfaction in, 74-79
    (*see also* Conflicts, Costs, Maps, Participation)
Review: *see* Technical review

Right-brain methods, 120-27
Roberts, S., 262
Ross, D., 6, 289
Round robin review, 234-35
Royalty payment example, 262-63
Rubin, M., 123
Rule of Three, 211

SADT™, 6, 289
Sample participation, 74-77
Satir, V., 146, 290
Satisfaction: *see* User satisfaction test
Schedule constraint, 181-82, 190-91, 193-200, 286
    meetings and, 86, 91, 112-14
    when do you need it? graph, 193
Schulman, E., 1, 286, 291
Shen, V., 286
Simile, 52-53, 55, 57
Simulation, 31
    computer, 77
    star, 22-33, 92
Sketching, 120-23, 127, 289
    Wiggle Charts and, 121-23
Social behavior, 130
Solution idea, 50-51, 55
Solution space, 173-85, 186-200
    for elevator example, 174, 179
    for Superchalk, 173-79, 186-88, 192
Spiral of Exploration, 105-106
Spradley, J., 78, 290
Standards, 180, 182-84
Star example, 22-33, 93-94, 97-101
Starting points, 49-58, 260, 277, 288
    universal, defined, 55
State space, 172-85
Stepwise refinement, 12
Strategic Advantage, 78
Straus, D., 89, 287
Suci, G., 289
Superchalk example, 47-48, 68, 147-48, 215-16
    attributes of, 161, 164-68
    constraints for, 171-79
    expectations and, 201, 205
    functions of, 149
    preferences of, 186-88, 192

solution space, 173-79, 186-88, 192
     testing for, 252
Surrogate participation, 74-78
Swedish army example, 9
Swiss Army knife-type system, 266
Swiss map example, 11
Swiss village example, 43

Tannenbaum, P., 289
Teakettle example, 152-55
Team-building activities, xvi, xvii, 139
Technical review, 199, 225-37, 287
     formal versus informal, 227-28, 230
     issues list, 231-32
     versus project review, 228-29
Technology idea, 51-52, 55
Testing, 215, 217, 219-24, 290
     ambiguity metric and, 219-22
     black box, 249-52, 257, 258
     constraints, 174-77, 185, 255-56
     preferences, 189
     satisfaction, 205, 212-13, 238-48
     technical reviews and, 199, 227-37
     test cases, 249-59
Tilt concept, 180
Time constraint: see Schedule constraint
Tools, requirements, 105-107
     (see also Brainstorm, Decision tree
     model, Heuristics, Interviewing
     techniques, Meetings, Right-brain
     methods, Technical review, User
     satisfaction test)
Trade-offs, 169, 195-200
Translation strategy, 11-13
Translucent function, 158
Transportation device example, 36-44
     ambiguity poll and, 42, 217-22
Turnpike effect, 272-73

Universal furniture design example,
     129-34
Universal starting point, 55
User:
     constituency, 70-79
     expectations of, 201-13
     -inclusion heuristic, 72-74
     -inclusion strategy, 73-74, 78-79

manual, 257
participation, 74-79
survey, 205
testing and, 252
-unfriendly features, 73-74
versus customer, 68-69
User satisfaction test, 205, 212-13,
     238-48, 289
     for Do Not Disturb Project, 239-40

Vanilla review, 233-34
Veblen's Principle, 71, 78
Videotaped interviews, 212
Visual tools, 120-27
von Neumann, J., xv, xvi, 199
Voting technique, 116-17

Walkover, example of, 225-27
Walkthrough, 225, 234, 235, 287
Warnier-Orr approach, 6, 289
Weinberg, D., 149, 291
Weinberg, G., 1, 34, 49, 69, 95, 122, 149,
     183, 191, 199, 225, 234, 249, 251, 283,
     290-91
Western Institute for Sofware Engineer-
     ing, 78
What if testing, 249-51, 258
What's it worth? graph, 192-97
     racing car example, 196
"What's it worth?" question, 60-61
Wheelchair example, 209
When-do-you-need-it? graph, 193-94
Wiggle Chart, 121-23, 290
Wilde, O., 277
Wish list, 161-66
Working title, 128-29
     for elevator example, 57
     for furniture design example, 132,
     137
Wright brothers example, 53-54
Wright, F., 121

Yourdon-Constantine approach, 6, 289

Zeroth Law of Product Development,
     198

MORE COMMENTS FROM THE REVIEWERS

"After reading this book, you'll never again trust dated requirements *— no matter who hands them to you, or how solemnly or frantically they are handed over. You will also be armed with some ways to discover the real requirements yourself."*

> — Ken de Lavigne, Senior Member of Staff
> IBM Quality Institute

"The title lays it out, that exploring requirements does imply quality before design, and the text provides the social, psychological, and intellectual processes to carry it out. Gause and Weinberg are unique in their experiences and abilities in the subject."

> — Harlan D. Mills, Professor
> Florida Institute of Technology

"long past due and sorely needed . . . the first book that describes how to negotiate the requirements of a product . . . to ensure a product of value."
> — Judy Noe, Partner
> The Strategic Advantage

"most timely as it focuses on the unambiguous approach to capturing customer requirements from a process viewpoint. . . . Through clear and telling examples, and using a straightforward, no-nonsense and easily readable style, Gause and Weinberg clearly show that the requirements process is not only a management tool but also the motor behind the entire enterprise of developing products."

> — Gabriel A. Pall, Vice President
> Juran Institute, Inc.